D1553619

THE GLOBAL RESTRUCTURING OF AGRO-FOOD SYSTEMS

EDITED BY

Philip McMichael

Cornell University Press

ITHACA AND LONDON

First published 1994 by Cornell University Press.

Printed in the United States of America

♾ The paper in this book meets the minimum requirements of the American National Standard for Information Sciences—Permanence of Paper for Printed Library Materials, ANSI Z39.48-1984.

Library of Congress Cataloging-in-Publication Data

The Global restructuring of agro-food systems / edited by Philip McMichael.
p. cm. — (Food systems and agrarian change)
Includes bibliographical references and index.
ISBN 0-8014-2940-4 (cloth). — ISBN 0-8014-8156-2 (paper)
1. Agricultural industries. 2. Food supply. I. McMichael, Philip. II. Series.
HD9000.5.G588 1994
338.1'9—dc20 93-48160

Contents

Preface

This book originated in a workshop held at the Rural Sociological Society (RSS) meetings in Columbus, Ohio, in the summer of 1991. Organized at the suggestion of Frederick Buttel, then president of the RSS, the workshop was successful in two respects: first, it generated considerable interest among the forty or so people who remained after the main conference, all scholars working in the area of the political economy of agriculture; and second, the papers presented converged thematically in several ways.

The presenters had been asked to address questions of food system and agrarian change in the late twentieth century. As the resultant chapters show, they invariably located these questions in the period of global restructuring from the 1970s to the early 1990s, with two consequences. Each chapter develops a part of a matrix of changes occurring in different regions or in different commodity complexes. This overarching theme of restructuring addresses a significant transition in the world order—and does so from the novel perspective of the transformation of agricultural and food systems. Consequently, this book locates these changes in the emergence of the twenty-first-century era—an era of a new global configuration that will be characterized more by the "food" and "green" questions than by the "agrarian" question per se.

The following individuals deserve my gratitude for their assistance at various stages of the book's development: Fred Buttel, for his inspiration to hold the workshop and his involvement as a discussant along with Alessandro Bonnano, Shelley Feldman, and Gary Green; the authors of the individual chapters for their participation in this project; Billie DeWalt, for his substantive contributions to the organization of this collection; Peter Agree, for his editorial encouragement and sound advice; Helene Maddux, for her editorial attentiveness and good humor; Tonya Cook and Margo Quinto, for their skillful copy editing; Philip Harms, for his timely assistance with preparation of the manuscripts; and Le Padgett, for her tireless work in tracking down authors and typing the essays.

Finally, I thank the *Third World Quarterly* for granting permission to reprint Harriet Friedmann's essay (Chapter 11), which first appeared in volume 13, issue 2 (1992), and *Policy Studies Review* for granting permission to reprint Kathleen Stanley's essay (Chapter 5), which is a revised version of a previously published article in volume 11, issue 2 (1992).

PHILIP MCMICHAEL

Ithaca, New York

Contributors

PASCAL BYÉ is a senior research fellow with the French National Institute for Agronomic Research (INRA), Department of Economics, in Montpelier. Specializing in the analysis of agriculture-industry relations, he has stressed the impact of innovations and new technologies on agro-food system and structure. His present program emphasizes the historical diversity of social, economic, and technological regimes.

MARIA FONTE is Associate Professor of Rural Sociology at the University of Naples, Italy. Her teaching and research topics include the structure of agriculture and the technological evolution in the agro-food system, with special emphasis on the social and economic consequences of modern application of biotechnology to agriculture.

WILLIAM H. FRIEDLAND is Professor Emeritus, having been Professor of Community Studies and Sociology for twenty-two years at the University of California, Santa Cruz. His research has concentrated on the social and technical organization of agriculture, and recently he has worked on the global fresh fruit and vegetable industry.

HARRIET FRIEDMANN is Professor of Sociology at the University of Toronto. Most recently, her research has taken her in two directions: cultural and material aspects of diets in relation to sustainable food economies, and the place of food within the Cold War and post–Cold War shifts in international power and property relations.

CHUL-KYOO KIM is a postdoctoral research fellow at the Center for Area Studies, Seoul National University. He is currently conducting research on East Asian development in the context of global restructuring.

GEOFFREY LAWRENCE is the Foundation Professor of Sociology at the University of Central Queensland in Rockhampton, Queensland, Australia. He is currently undertaking a major study into the social impacts of agricultural biotechnological development.

LUIS LLAMBI is Professor of Anthropology at the Venezuelan Institute for Scientific Research (IVIC), Caracas. His research focuses on comparative agrarian structures in Latin America and the global political economy of agro-food systems.

PHILIP MCMICHAEL is Associate Professor of Rural and Development Sociology and Director of the International Political Economy Program at Cornell University. His research interests include historical sociology of the world economy, global regulation, and current agro-food system restructuring in the Pacific Rim.

TERRY K. MARSDEN is Professor of Geography and Earth Resources at the University of Hull, Great Britain. He has conducted extensive research on rural systems and agricultural change in Britain.

DAVID MYHRE is a visiting research fellow at the Center for U.S.-Mexican Studies at the University of California, San Diego. He is Coordinator of the Center's Ejido Reform Research Project, a three-year study of the consequences for rural Mexico of recent reforms to agrarian laws and agricultural policies. His research interests center on rural social movements, the politics of agricultural finance and investment, and the restructuring of state-peasant relations.

LAURA T. RAYNOLDS is Visiting Assistant Professor of Sociology at the State University of New York, Binghamton. Her research focuses on current processes of restructuring in Latin America and the Caribbean, linking global transformations, shifting state policies, patterns of firm reorganization, and the changing livelihoods of local populations.

KATHLEEN STANLEY teaches sociology at Western Oregon State College. She has worked on the problem of working households in the world economy, and current research focuses on migrant labor in the Willamette Valley of Oregon.

DAVID VAIL is Adams Catlin Professor of Economics at Bowdoin College. He is currently an associate at the Swedish University of Agricultural Sciences Information Center for Rural Planning and a board member of Stockholm University's Swedish Program. His research has centered on the greening of agricultural policy in advanced industrial societies.

Frank Vanclay lectures in sociology at Charles Sturt University and is a key researcher with the Centre for Rural Social Research at Charles Sturt University in Wagga Wagga, New South Wales, Australia. He has undertaken numerous consultancies for government departments relating to social and environmental issues in agriculture.

Sarah Whatmore is Reader in Human Geography at the University of Bristol. She has conducted extensive research on the political economy of agriculture and rural land-use change, particularly its gender and environmental dimensions. She currently holds a Research Fellowship under the Global Environment Change program of the Economic and Social Research Council in Britain.

The Global Restructuring of Agro-Food Systems

Introduction: Agro-Food System Restructuring—Unity in Diversity

Philip McMichael

The late twentieth century is evidently a transitional period, possibly the end of an era. Widespread talk of restructuring—of economies, political systems, states and empires, the world order—attests to this new threshold. How to interpret this twenty-first century threshold, its origin and direction, is problematic. Some perceive current events in the short term (the end of U.S. hegemony, national regulation, the Cold War, and the Third World), others in the medium term (the collapse of the Second World and Communism), and others in the long term (the denouement of nineteenth-century movements of democracy, nationalism, and socialism and global environmental deterioration). However these events are viewed, they raise the question, Are these restructuring processes presaging a new order or simply rearranging the old order, now in crisis? This collection of essays directly addresses this question from a fresh perspective: that of changes in food, and agrarian, systems. Although commodity, sector, and place vary among the essays, the authors situate their inquiries in relation to a common transition.

The common transition stems from the crisis of the post–World War II political-economic order, otherwise known as the *Pax Americana*. The crisis of the dollar as international reserve currency led to the dismantling of the Bretton Woods system in the early 1970s (Block 1977). Bretton Woods, institutionalized in stable currency exchanges supported by the International Monetary Fund (IMF) and the World Bank, and underwritten by U.S. deficits, framed the postwar period. This was a time when Keynesian countercyclical economic policy sustained national systems of regulated capitalism. This institutional nexus of international and national political economy was most fully elaborated in the metropolitan states. It anchored a nationally managed, high-wage social contract between capital and labor characterized by "a certain set of labour control practices, technological mixes, consumption habits, and configurations of political-economic power . . . called Fordist-Keynesian" (Cox 1987:124).

This combined form of national and international regulation Ruggie terms "embedded liberalism," where "multilateralism would be predicated upon domestic interventionism" (1982:393). The national model of economic growth, an *ideal* representation of U.S. political economy institutionalized in the Bretton Woods system, was diffused across the expanding state system in the context of decolonization, informing development strategies of import-substitution in both industry and agriculture (Friedmann and McMichael 1989). The model legitimized protection of national economic sectors, based on key farm and labor constituencies of the national polity sustained through Keynesian fiscal policy.

This form of national capitalism, however, was relevant only so long as the dollar/gold standard operated, linking currency stability to stability in national balances of payments (Phillips 1977). After U.S. deficits grew unmanageable by the late 1960s (Kolko 1988), with rising domestic wage demands and growing transnational movements of capital (Arrighi 1982, Lipietz 1987), the U.S. government uncoupled the dollar from gold in 1971. A new era of unstable trade and floating exchange rates began, undermining national projects and elevating internationalization forces. The formerly stable interstate hierarchy (within the north and between north and south) unraveled, and international capital markets, based initially in Eurodollar markets in the 1960s and universalized in the 1970s via the prosperity of the oil-producing states, came into their own. They formed, for the first time, "a single world market for money and credit supply" (Harvey 1989:161). In the absence of a stable system of (national/international) regulation, nation-states have been set adrift to negotiate their own competitive position in the world economy (Cerny 1991). This negotiation process, across two phases of relatively unlimited international liquidity (the 1970s) and relatively limited international liquidity (1980s monetarism and the debt regime), has enhanced international restructuring forces within, and among, states. The result has been the reorganization of the institutional and production and exchange relations of capitalism.

These phases correspond to the common time frames of the various instances of agro-food system restructuring analyzed in this book. Although temporal patterns link and define the cases, there are variations in space. And the Australian case is perhaps an exception that proves the temporal rule. Australian economic policy rationalization preceded the neoliberal movement of the 1980s by a decade, no doubt legitimizing its subsequent role as host of the Cairns Group.[1] Australia lost its traditional markets in the 1960s when Britain joined Europe's Common Agricultural Policy. In reorienting its markets and politics in accordance with the United States' new Pacific Rim strategy, Australia began a profound, and early, process of economic restructuring (see Chapter 3), reflecting the prior organization of national economy around a trade dependence on imperial preference.

A particular reconfiguration of foreign markets preceded the Australian shift from nationally coherent to globally competitive economy, but the general *temporal* pattern was the restructuring of global markets (1980s) following the breakdown of institutions of national coordination (1970s). This period of crisis and restructuring frames the analyses presented here. On another axis, there was also a profound shift in *spatial* relations governing the timing of the decline in national economic organization. Soviet grain purchases of American grain in the early 1970s (inflating food prices, expanding farm surpluses, and further destabilizing international trade) began the informal dissolution of the Cold War bloc structure, which underpinned and demarcated the stable Bretton Woods order within which national capitalism (and farm programs) flourished (Friedmann 1993). Relations of time and space are interlinked in the restructuring process.

Though the authors investigate a variety of forms (organizational dimensions of agro-food systems as political and technological constructs) in a variety of spaces (states, sectors, and world regions), time is a unifying coordinate. In fact, the unity in time is another expression of the universality of restructuring, which, in some instances, standardizes processes across space and in other instances reconfigures spatial relations as differentiated elements of a common process.

There are two distinguishing features of this collection. First, individually the essays are up-to-date treatments of agro-food system restructuring. Second, together they provide a multidimensional and comparative treatment of the subject, revealing a unity in the diversity of restructuring processes. The obvious material diversity among agri-

[1] The Cairns Group includes the agricultural exporters Argentina, Australia, Brazil, Canada, Chile, Columbia, Fiji, Hungary, Indonesia, Malaysia, New Zealand, the Philippines, Thailand, and Uruguay and espouses elimination of trade-distorting farm protectionism.

cultural and food subsectors is represented in analyses of three key arenas of restructuring: grains (wheat and rice), livestock, and fruits and vegetables. There is also social/organizational diversity across space, addressed in the variety of northern and southern states or regions represented here, from Sweden to Mexico. The unity derives from the interconnectedness of the restructuring process, however distinctive the local form. The production and circulation of agricultural commodities increasingly shares common financial and technological relations, as global capital markets and transnational firms integrate, and transform, food systems (e.g., Chapters 4, 7, and 10). Social diets, increasingly standardized across space and culture, but not necessarily across class, reflect and condition the global reorganization of agriculture (Chapters 1, 7, and 11). This influence is exemplified in regional reorganizations of livestock production in the United States (Chapter 5) and Australia (Chapter 3) and emerging nontraditional agro-export systems of fruits and vegetables in the Caribbean (Chapter 9) and Latin America (Chapter 8).

Together, the essays present three key aspects of restructuring: (1) questions about its status (its manifestations, meaning, origin, and direction); (2) its dimensions (financial, productive, technological, cultural, political, and organizational); and (3) its locations (differentiated by time: postcolonial, post–Keynesian/Fordist, post–Bretton Woods, and so on; and differentiated by space: metropolitan/peripheral, global/regional, cross-sectoral/subsectoral, and the like). The diversity of forms and contexts in which agro-food restructuring is expressed demonstrates at once its complexity and its universality. It is universal not simply because agricultural and food systems are being transformed globally, but also because this transformation implicates industrial, financial, and service sectors. Arguably, the very definition and content of current restructuring involve a fundamental process of intersectoral integration, unconstrained by national boundaries. Accordingly, restructuring redraws economic and political boundaries and transforms the spatial categories with which social scientists work. For these reasons, the essays in this volume offer fruitful lines of inquiry into current social change.

One of the dominant themes unifying the essays is how agro-food systems are undergoing restructuring in an era of declining national regulation. Insofar as the regulatory capacities of nation-states have declined and their goals have shifted, markets and technologies have been reorganized considerably. Flexibility in production and marketing strategies, noted particularly in industry (Piore and Sabel 1984, Hirst and Zeitlin 1990), has emerged also in agro-food systems. Growing specialization in food systems is apparent in the global expansion of niche markets for fresh and processed foods. Agribusiness corporations have revitalized contract farming on a global scale (Watts 1990), and transnational fresh fruit and new vegetable corporate distributors service proliferating metropolitan markets (see Chapter 7). Developments in intensive meat complexes display this new combination of flexibly based mass production (see Chapters 3 and 5).

These developments exert pressure on states to pursue national competitiveness rather than national coherence with respect to their agricultural sectors (McMichael 1992). In the meantime, agricultural protectionism in the north continues, intensifying the competition for markets in which to dispose of farm surpluses. In this context, states as well as agribusinesses have been involved directly in the restructuring of the world food economy, in part through the General Agreement on Tariffs and Trade (GATT) negotiations. "Food security" politics informs national and international debates, interwoven ambiguously with demands for agricultural liberalization in a proposed world free trade regime (see Chapters 1, 2, and 6).

A variety of issues within this postnational regulation theme shed light on the issue of transition:

- Is restructuring a breakthrough or simply a response to the breakdown of the era of national regulation?
- What are the key agents of change?
- How are the new forms of regulation expressed in states, labor markets, financial systems, commodity markets?
- How do particular states, producers, and firms contribute or respond to restructuring?
- What does the international division of labor in the post–Bretton Woods era look like?

Answers to these questions lie in the individual and cumulative arguments of the authors as they locate and interpret their respective topics. Authors address the issues of restructuring that animate the contemporary literature, such as post-Fordism, global regulation, new social movements, environmental imperialism, the new international division of labor, market and labor force segmentation, and transnationalization. Perhaps the framing issue is that of the retreat from Fordism, since it expresses the decline of the politically managed model of mass production upon which postwar national prosperities were based (e.g., Marsden 1992).

The conditions for the Fordist system stemmed from the consolidation of monopoly capitalism, which encouraged the perpetual reorganization of the labor and production processes (Tickell and Peck 1992). The associated expansion of productivity underwrote a high-wage national regulation economy in which states managed capital investment and maintained high corporate and individual consumption through Keynesian fiscal policy and trade protection. Long runs of standardized, durable commodities (including food) characterized this regime of accumulation (Aglietta 1979), which exhausted itself in the late 1960s through rising consumer expectations and growing resistance of organized labor to the regimented workplace (Arrighi 1982, O'Connor 1986). The unraveling of this order, signaled by the end of the Bretton Woods system, generated new decentralized forms of social economy and new regulatory mechanisms. In the wake of the diminishing regulatory capacities of national states, for example, the IMF has assumed new powers of international regulation, beyond its initial role of redistributing liquidity (Wood 1986).

Whether these new forms of social economy can be termed post-Fordist is a matter of intense debate (Hirst and Zeitlin 1990); that is, are they symptoms of crisis (Gordon 1988, Clarke 1990) or indeed the shape of the future (Harvey 1989)? Clearly, forms of mass production continue, but Fordism yields to Toyotism (lean production) as new forms of flexible organization of the labor force and new specialty-product marketing complement or compete with the mass economies established within the framework of the postwar Keynesian state. At the same time, however, it appears that economic management may return to favor, at least for metropolitan states, as they confront rapid global economic and social change and move to secure strategic technologies and infrastructures.

Related to the retreat from Fordism was the relocation of low-skill manufacturing processes to the south in the 1970s, giving rise to the concept of the "new international division of labor" (Fröbel, Heinrichs, and Kreye 1989). That term has since been extended to the agricultural sector (Sanderson 1985), where southern agricultures have been transformed by the relocation of some labor-intensive production alongside the capital-intensive (mainly grain) sectors (DeWalt 1985, Llambi 1988) and new agro-exports have appeared alongside traditional tropical exports (Raynolds et al. 1993).

In this context of rapid, but uneven, growth in southern economies, fueled by new offshore, international capital markets, the regulated high-wage economy associated with the Keynesian-Fordist project is no longer either possible or adequate. New corporate strategies have emerged, geared to global and regional, rather than national, markets for new specialty products (beyond mass products) oriented to segmented niche markets, such as those for fresh, exotic, and ethnic foods. Just as product markets, whether local or foreign, increasingly conform to global standards, production and distribution systems increasingly combine local and transnational capital in joint ventures, leading one commentator to reconceptualize these new forms of economy as "transnational practices" (Sklair 1991).

In sum, each case presented here offers an individual lens on a general synchronic trend in political-economic reorganization. In particular, the early to mid-1970s are universally identified as the key threshold of collapse of the postwar national regulatory order (e.g., fixed dollar exchange rates, managed trade, national farm programs, mass production, welfarism). Within the next decade, new, global forms of political economy emerged (e.g., worldwide securitization, the debt regime and southern agro-exporting, specialty production and niche marketing, the GATT Uruguay Round and agricultural liberalization, corporate strategies of global sourcing and distribution).

The form of restructuring is governed by time/space relations. These locational coordinates condition the particular sociopolitical relations of agro-food system restructuring, subdivided in this book into three dimensions: states, sectors, and global regions. Of course each dimension is in a sense intrinsic to the other, and one thinks initially of the Russian nested-doll metaphor as a means by which to conceptualize abstractly this relation. However, since the fit is by no means uniform, precisely because of the uneven relations of form, space, and even time within and across these arenas, the process must be understood as heterogeneous and unpredictable. This aspect is addressed in Part IV of the book, "Common, Contradictory, and Contingent Forces," which offers three global overviews that implicate the specific cases discussed in the first three sections on states, sectors, and regions.

State-Level Restructuring

A key issue of state-level restructuring is the crisis in unregulated, and cutthroat, agricultural trade rivalry stemming from the managed surpluses of the metropolitan farm sectors. This crisis underlies the new focus on farm programs in the GATT negotiations, which have provided a pretext, if not an institutional context yet, for agricultural liberalization (McMichael 1993). Agricultural liberalization has different consequences and forms for the various domestic farm and food security programs across the state system. This collection looks at three cases: the advanced form of deregulation in Australia leading to an alternative agro-industrial strategy; the erosion of Swedish corporatist farm policy in the context of the crisis of the model of intensive, specialized farming and European Community expectations; and Japanese and South Korean recalcitrance in the face of U.S. bilateral pressure on highly protected East Asian rice subsectors to relieve its agricultural surpluses.

Common to these particular cases is that globalization processes crystallize national policies, forcing a reexamination of their salience in a changing world order. The demise or redundance of national regulatory systems in agriculture is being played out dramatically in Japan and South Korea, the subject of Chapter 1, by Philip McMichael and Chul-Kyoo Kim. From the U.S./GATT perspective of trade liberalization, each state's industrial export strategy is compromised by farm protectionism, and yet the latter is deeply institutionalized in national politics and ideology—especially with respect to rice self-sufficiency. McMichael and Kim examine the complexity of this issue through comparative analysis of the evolution of these East Asian agricultural systems: where rice farming symbolizes national economy at the same time as, and in part because, complementary foodstuffs (such as flour and meat products) have been increasingly available via U.S. agricultural exports. Two factors intertwine in the politics of East Asian liberalization. The relevance of rice protectionism is questionable when Japanese and South Korean agricultural systems are increasingly internationalized anyway; and an intensified trade rivalry as "food power" has become consequential to the U.S. export strategy, especially given the East Asian industrial export offensive. In sum, the inseparability of national and international relations is made particularly evident in this era of declining national regulation.

In Chapter 2, David Vail extends the emphasis on the domestic forces involved in negotiating globalization pressures. He examines the dismantling of Sweden's "negotiated economy" as a realignment anticipating agricultural reforms in the EC's Common Agricultural Policy (CAP), as well as the liberal sentiments driving the GATT's Uruguay Round. In addition, Vail stresses the demise of the model of agriculture associated with European systems of national regulation—one in which the limits of the agro-industrial model's chronic surpluses and environmental degradation have increasingly undermined institutionalized support for the farm sector (based on a farm-labor alliance founded in postwar food security policies). Although the GATT and CAP negotiations remain unresolved, prospective membership in the EC will reinforce the dismantling of the Swedish form of national corporatism, though, ironically, agriculture remains one of the most corporatist aspects of the EC. Swedish agricultural and food policy points toward a phased reconfiguration of farm supports via the elimination of price supports, with income compensation to disadvantaged regions and for "greening" projects.

Geoffrey Lawrence and Frank Vanclay, in Chapter 3, examine the establishment of a new feedlot industry in Australia's Murray-Darling Basin. They document this development as both indicative of the government's general deregulation of agriculture (see Lawrence 1987) and a reflection of the world economic position of Australia, in two senses. First, the restructuring of Australian agriculture is geared to stemming economic deterioration (deindustrialization) via intensification of high-value agricultural exports such as beef to the expanding East Asian markets. Second, the irony of this strategy (Australia as a food factory financed with Japanese capital) is that it promotes what Lawrence and Vanclay provocatively term "environmental imperialism," where ecologically damaging feedlot production associated with metropolitan methods of mass production (Fordism) is moved to nonmetropolitan regions.

Sectoral Restructuring

The further erosion of sectoral boundaries under general economic restructuring, such as integration of agro-food systems into financial and corporate strategies of value adding (Chapter 4), is manifested in the related processes of recomposition of the labor force in agro-industries, such as meatpacking (Chapter 5), and the defensive mobilization of small farmers caught in the trap of establishing economic viability through further financial incorporation (such as the treatment of Mexican *campesinos* in Chapter 6).

Terry Marsden and Sarah Whatmore situate agro-food system and regional developments within the process of financial reorganization in Chapter 4. Their goal is to distinguish and locate the food system within broader economic restructuring. Changes in British financial markets toward greater capital mobility (e.g., the rise of securitization—requiring external funds and thereby encouraging internationalization of finance, especially vis-à-vis the food sector) have increased operational flexibility. Realignments between financial and industrial capital encourage restructuring (such as mergers) in the food sector, especially corporate concentration. The theoretical point is that corporate concentration, which is related to the inelasticity of food demand and corresponding growth strategies of value adding, lowers barriers to industrial capital investment in farming. This concentration has two major effects: (1) it leads to multiple banking (the fragmentation of corporate financing among several competing banks) and the proliferation of financial product services (especially as public consumption is privatized), and (2) it generates new flows of capital and technology to food producers and regions.

Financial restructuring and responses to the 1980s farm crisis vary among nations. Compared with European systems of state-sponsored agricultural finance, British financial deregulation hastened the process of agro-industrial concentration and financial penetration of the farm sector. Marsden and Whatmore argue that the British experience shows the way for European states, which converge as the farm protection system is reformed. In a context of rising farm debt, and declining land (collateral) value, banks have moved to subordinate producers to more efficient operation, including control over inputs via credit packages. They conclude that land-based agriculture is yet another sector to which the synergies of industrial and financial capital apply, as banks collaborate with, and depend upon, industrial capital in a highly contingent and competitive process.

Recent developments in intensive meat complexes involve the new forms of flexibly based mass production associated with industrial restructuring in general. Beef production is restructuring globally, as massive new feedlots are located near feed-supply regions and sources of cheap meatpacking labor. In Chapter 5, Kathleen Stanley shows how U.S. mass-production meatpacking plants have been reconstituted, driven by new marketing strategies (for meat products such as deboned and boxed meats for domestic and foreign markets) and intensified competition. The new plants involve more complex processing of meat products requiring greater automation and are located in the sparsely populated High Plains near the huge feedlots associated with newly specialized feed grain producers. Facilitating this restructuring is a growing supply of cheap immigrant, and U.S.-born migrant, labor to fill the peripheral high-turnover jobs that complement the stable core of jobs that have not been de-skilled via automation of the livestock "disassembly" line.

Discussion of restructuring would not be complete without an analysis of how rural social movements respond to, and indeed reformulate, the effects of global forces within the agrarian sector. David Myhre's chapter (6) examines a contemporary Mexican peasant movement to manage the rural credit system in a context where international agencies have pressured the state to reduce its social expenditures, including withdrawal from historical support for rural institutions (such as the ejidos). UNORCA, the peasant coalition formed in 1985, has been pressing for decentralized control of the rural economy (requiring certain financial support) just as the state has been responding to global pressures to cede control of economic institutions (such as rural credit) to private financial interests. Since the latter apply criteria of international competitiveness even to small-scale peasant projects, the peasant credit union movement is faced with reformulating their politics to negotiate the challenges of globalization.

Global Region Restructuring

For forms of regional restructuring, there is the process of export substitutionism, organized jointly by transnational firms and governments to capture world market niches that emerged in the context of the debt regime of the 1980s, especially in Latin America. Chapters 7, 8, and 9 analyze the phenomenon of nontraditional exports from the south, such as fruit and vegetables (fresh or processed), from three distinct angles: corporate organization in the distribution sphere (Chapter 7), the limits and possibilities of Latin American state export strategies in this new international division of labor (Chapter 8), and an extended example of some of these limits and possibilities associated with the restructuring of export agriculture in the Dominican Republic, emphasizing the instability of this strategy for southern states (Chapter 9).

In Chapter 8, William Friedland addresses the relatively recent globalization of the fresh produce (other than bananas) segment of the food industry, whose global reach began in the colonial era. Global organizational strategies in the fresh fruit and vegetable system occur within the distribution sphere, mediating producers and marketers, which remain locally or nationally based. Friedland compares two distinct (but both current) corporate strategies exemplified by (1) firms such as Dole, Chiquita, and Del Monte Tropical, which have a historical base in banana production and extend into other commodities, and (2) new firms such as Albert Fisher and Polly Peck International, which assemble capital to acquire firms with experience in fresh produce distribution and then develop the logistics to operate transnationally. For Friedland, globalization is facilitated by the concentration of food retailing in metropolitan markets, and driven by metropolitan consumer expectations of year-round fresh produce and a variety of tropical products.

Luis Llambi, in Chapter 8, examines the context within which southern countries have been promoting higher-value nontraditional exports of fruit and vegetables to replace traditional tropical exports associated with the colonial era, on the one hand, and to meet the requirements of debt service on the other hand. Using the cases of Chile, Brazil, Mexico, and Colombia in particular, he considers the global and local conditions for such agricultural restructuring, developing some working hypotheses about the efficacy of the new Latin American agro-export strategy. Although noting the real opportunities in the new niche markets associated with new (post-Fordist) patterns of metropolitan consumption, Llambi is less than sanguine about the long-term economic possibilities (a renewed "colonial" division of labor?) as well as the fragility of current international trade dynamics.

The endemic instabilities of the world market for southern exports becomes a central theme in Laura Raynolds's investigation of the Dominican state's attempts to restructure export agriculture in Chapter 9. This theme has two sides: first, the historical decline of revenues from traditional agricultural exports, and second, pressure from international institutions for liberalization via structural adjustment—both of which lead to export-substitution strategies (such as nontraditional exports). In detailing shifts in state policy and the characteristics of the commodities and firms involved, Raynolds identifies a central paradox of this strategy of liberalization. That is, in order to work, the strategy requires local regulation to stabilize the conditions for accumulation (which in turn would reduce cross-national capital mobility), as well as global regulation to reduce metropolitan protectionism that threatens agro-export substitution. Thus restructuring at the global region level is very much a process negotiated by states. However, where these states, such as the Dominican Republic, lack significant domestic markets to absorb risk, they find their fortunes governed by their competitive position within the new exporting region.

Common, Contradictory, and Contingent Forces

The several levels of restructuring share common elements such as technology, international food complexes, and institutional forces. Byé and Fonte (Chapter 10) argue that technology unevenly combines residual and emergent forces within and among different subsectors and that it is unpredictable in the direction of its application and consequences. Friedmann's chapter analyzes the international food complexes of wheat, durable food, and livestock. Institutional forces—global regulatory mechanisms such as the IMF and the GATT—are dealt with in the concluding chapter. There is a complex, nonlinear, and politically contingent aspect to all these different forces unifying different aspects of agro-food systems across the world. Yet we can develop some analytical perspectives to bring these changes into relation to one another so that some patterns of change, as well as political opportunities for change, can be identified.

In Chapter 10, Pascal Byé and Maria Fonte offer a provocative view of the technical dimension of agro-food system restructuring, with an unconventional conception of agricultural revolution. The authors propose that we think about agricultural technical change in a nonlinear fashion—going beyond the issue of the pace of technical change. Conventional speculation about the likely impact of biotechnologies is essentially quantitative, they claim, when the real issue is a qualitative one. It concerns the scope of technical change, which in turn depends on the mutual conditioning of prior technologies, corporate structures and strategies, and reregulation policies reflecting environmental, market, and consumer concerns. Their theme is twofold: (1) given the context (investment and technological inertia, uneven production and circulation conditions, land and commodity pricing policy, etc.), new agricultural technologies such as biotechnology do not in themselves constitute revolutionary forces because they do not emerge sui generis—they are conservative or transformative forces depending on sectoral location and other circumstances; and (2) related to this consideration, agro-food producers may revolutionize agriculture to the extent that they successfully adapt to new social definitions of food, agriculture, and rural environments.

The complexity and diversity of modern agriculture forbid linear scenarios of agro-industrial homogenization, in which agriculture becomes merely a subsector of industry. In fact, the relative flexibility and size of biochemical corporations give them an edge as vehicles of technical change, precisely because they are in the best position to bridge both mass and specialized forms of production. Ultimately, Byé and Fonte are claiming that the restructuring of agro-food technologies in the current juncture combines residual (mechanical logic) and emergent (biochemistry) technologies in uneven responses to the exigencies of the accumulation crisis.

In Chapter 11, Harriet Friedmann analyzes three key segments of the world food economy since World War II: the wheat, durable foods, and livestock complexes. Each complex has definite political origins and spatial consequences. The wheat complex, arising out of managed surpluses in metropolitan farm programs, effectively reorganized north/south relations by institutionalizing southern grain dependency—first, through the U.S. food aid regime, which undermined local agricultures, then through the market as southern states came to depend on the cheap grains dumped on the world market through U.S./EC export rivalry. The unmanaged trade arising out of this rivalry underlies the crisis of this food regime and current GATT attempts to reregulate agricultural trade. The durable foods complex, based in new postwar food manufacturing techniques, altered the terms of trade for tropical products through metropolitan substitution, especially those manufactured fats and sweeteners that displaced tropical oils and sugar. In this context, traditional southern trade complementarity has yielded to competitive production of nontraditional exports in the south. Finally, the livestock complex, integrating feed producers with feedlot technology across national boundaries, has been the chief vehicle of global restructuring of agricultures, as affluent states (in north and south) have developed their own livestock industry, and other southern states have become significant feed exporters, rivaling U.S. dominance in world food markets. This transformation has set the stage for the growing displacement of food crops by feed crops (exacerbating southern wheat dependence) and for the international trade in cereals (embodying generic ingredients, such as starch and sweeteners) to anchor global sourcing strategies of transnational companies for the new inputs of an exploding food manufacturing sector.

The concluding chapter builds on the themes of the preceding chapters. It has two goals: to identify new trends in the world food economy and to suggest new lines of inquiry for social scientists interested in world agriculture and food systems and their particular, and significant, role in the future world order. Lengthy maneuvering over the GATT Uruguay Round negotiations, including the appearance of regional trading blocs, crystallizes some of the ongoing reorganization of the world food economy. This has two main sources. First, the breakdown of the managed trade associated with the Bretton Woods system and the shift of initiative away from states to banks in underwriting agro-industrialization geared increasingly to globalized markets, integrated by transnational food companies, and supplying new social diets differentiated by affluent consumption of specialty foods and mass consumption of manufactured foods. Second, monetarism and the debt regime in the 1980s have wrought considerable change in the composition and policy of southern states, threatening local food security and encouraging new agro-export strategies to service debt and retain international credit-worthiness. On top of all this, transnational corporations would deploy GATT to dismantle universally farm protection (an institutional barrier to their global operations), accelerating the commodification of food and the restructuring of metropolitan farming. All these institutional changes are important foci for future research. In addition, the new dynamics of the world food order warrant investigation; with growing southern rivalry to the United States, the breakup of the Soviet bloc, the Japanese import complex, regional trading blocs, and so on. The key will be situating these changes in relation to one another, a task modeled by this collection.

For agro-food systems, the pace and scope of changes occurring at this time are phenomenal. Not only are food cultures and technologies undergoing extensive restructuring, but also national farm sectors, and indeed nation-states, are being redefined as the world integrates economically and socially. The content of contemporary agrarian politics is inexorably shifting from the (nationalist) agrarian question to the (internationalist) food and green questions, as late twentieth century polities and cultures confront accelerating reconfiguration of agriculture and natural resources (Goodman and Redclift 1991). This collection of essays ranges across the multidimensionality of this event, offering one fruitful glimpse of the unity in the diversity of restructuring processes.

References

Aglietta, Michel. 1979. *A Theory of Capitalist Regulation: The U.S. Experience.* London: New Left Books.

Arrighi, Giovanni. 1982. "A Crisis of Hegemony." In *Dynamics of Global Crisis,* Samir Amin, Giovanni Arrighi, Andre Gunder Frank, and Immanuel Wallerstein, pp. 55–109. New York: Monthly Review Press.

Block, Fred L. 1977. *The Roots of International Disorder: A Study of United States International Monetary Policy from World War II to the Present.* Berkeley: University of California Press.

Cerny, Philip G. 1991. "The Limits of Deregulation: Transnational Interpenetration and Policy Change." *European Journal of Political Research* 19: 173–96.

Clarke, Simon. 1990. "The Crisis of Fordism or the Crisis of Social Democracy?" *Telos* 83: 17–98.

Cox, Robert W. 1987. *Production, Power, and World Order: Social Forces in the Making of History.* New York: Columbia University Press.
DeWalt, Billie. 1985. "Mexico's Second Green Revolution: Food for Feed." *Mexican Studies/Estudios Mexicanos* 1: 29–60.
Friedmann, Harriet. 1993. "The Political Economy of Food: A Global Crisis." *New Left Review* 197: 29–57.
Friedmann, Harriet, and Philip McMichael. 1989. "Agriculture and the State System: The Rise and Decline of National Agricultures, 1870 to the Present." *Sociologia Ruralis* 29: 93–117.
Fröbel, Folker, Jurgen Heinrichs, and Otto Kreye. 1989. *The New International Division of Labor.* Cambridge: Cambridge University Press.
Goodman, David, and Michael Redclift. 1991. *Refashioning Nature: Food, Ecology, and Culture.* London: Routledge.
Gordon, David. 1988. "The Global Economy: New Edifice or Crumbling Foundations?" *New Left Review* 168: 24–64.
Harvey, David. 1989. *The Condition of Postmodernity.* Oxford: Basil Blackwell.
Hirst, Paul, and Jonathan Zeitlin. 1990. "Flexible Specialization versus Post-Fordism: Theory, Evidence, and Policy Implications." *Economy and Society* 2(1): 1–56.
Kolko, Joyce. 1988. *Restructuring the World Economy.* New York: Pantheon.
Lawrence, Geoffrey. 1987. *Capitalism and the Countryside: The Rural Crisis in Australia.* Sydney: Pluto Press.
Lipietz, Alain. 1987. *Mirages and Miracles: The Crisis of Global Fordism.* London: Verso.
Llambi, Luis. 1988. "The Emergence of Capitalized Family Farms in Latin America." *Comparative Studies in Society and History* 31: 745–74.
McMichael, Philip. 1992. "National/International Tensions in the World Food Order: Contours of a New Food Regime?" *Sociological Perspectives* 35(2): 343–65.
——. 1993. "World Food System Restructuring under a GATT Regime." *Political Geography* 12(3): 198–214.
Marsden, Terry. 1992. "Exploring a Rural Sociology for the Fordist Transition: Incorporating Social Relations into Economic Restructuring." *Sociologia Ruralis* 32(2/3): 209–30.
O'Connor, James. 1986. *Accumulation Crisis.* London: Basil Blackwell.
Phillips, Anne. 1977. "The Concept of 'Development.'" *Review of African Political Economy* 8: 7–20.
Piore, Michael, and Charles Sabel. 1984. *The Second Industrial Divide.* New York: Basic Books.
Raynolds, Laura, David Myhre, Philip McMichael, Viviana Carro-Figueroa, and Frederick H. Buttel. 1993. "The 'New Internationalization of Agriculture': A Reformulation." *World Development* 21(7): 1101–121.
Ruggie, John Gerard. 1982. "International Regimes, Transactions, and Change: Embedded Liberalism in the Postwar Economic Order." *International Organization* 36: 379–415.
Sanderson, Steven. 1985. "The 'New' Internationalization of Agriculture in the Americas." In *The Americas in the New International Division of Labor,* ed. Steven Sanderson, pp. 46–68. New York: Holmes and Meier.

Sklair, Leslie. 1991. *Sociology of the Global System.* Baltimore: Johns Hopkins University Press.

Tickell, Adam, and Jamie A. Peck. 1992. "Accumulation, Regulation and the Geographies of Post-Fordism: Missing Links in Regulationist Research." *Progress in Human Geography* 16(2): 190–218.

Watts, Michael. 1990. "Peasants under Contract: Agro-Food Complexes in the Third World." In *The Food Question: Profits versus People?* ed. Henry Bernstein, Ben Crow, Maureen Mackintosh, and Charlotte Martin, pp. 149–62. New York: Monthly Review Press.

Wood, Robert E. 1986. *From Marshall Plan to Debt Crisis: Foreign Aid and Development Choices in the World Economy.* Berkeley: University of California Press.

PART I

STATE-LEVEL RESTRUCTURING

1

Japanese and South Korean Agricultural Restructuring in Comparative and Global Perspective

Philip McMichael and Chul-Kyoo Kim

The current contention over the question of liberalization of the Japanese and South Korean agricultural sectors is a dramatic part of the story of global restructuring. It includes a range of issues framed by an elemental tension between national and international forces. On the national side, there are issues of food security for Japan and South Korea, the stability of their rural population (as farmers and political constituencies), environmental protection, the question of agricultural productivity, high domestic food prices (and therefore wage costs), and the growing fiscal consequences of expensive farm subsidies. On the international side, there are issues of trade (including Japanese and South Korean manufacturing export access to foreign markets, especially in the United States and the European Community), geopolitical pressures from the United States—anxious to secure its competitive advantage in farm products in the premier food-importing region—and, ultimately, the credibility of the Japanese in international forums such as the GATT. This tension is as yet unresolved, in large part because it is more complex than simply whether it makes economic sense for these East Asian states to liberalize their farm sectors. Indeed, the issue is universal: how should states adjust to a postnational regulation era?

Our premise is that the demise of the Bretton Woods system in the early 1970s marked the decline of an era of national economic regula-

We thank Alessandro Bonanno, Larry Burmeister, Ray Jussaume, Jr., and especially Mark Selden for their constructive suggestions on earlier versions of this essay. Philip McMichael also thanks Bruce Koppel and the East-West Center for research support.

tion, leading to a destabilizing increase of capital mobility on a global scale. In the absence of an ordered system of currency exchange and capital controls, based in the gold/dollar standard, nation-states have been compelled to renegotiate their position in a changing world economy. They have resorted to such measures as export initiatives, wage reduction, devaluation, and redistribution of public expenditures and credit facilities. This national renegotiation parodoxically elevates internationalizing forces within and among states as they strive to make the transition from national economic coherence to national economic competitiveness (see Cerny 1991, McMichael and Myhre 1991).

It is this paradox that frames the question of agricultural liberalization in East Asia, where mature Japanese and South Korean industrial export sectors coexist with small-scale and highly protected farm sectors. A critical issue for these East Asian states is that, from the U.S./GATT perspective, the future of their industrial exports is compromised by farm protectionism, as it both limits export markets for surplus agricultural producers such as the United States and threatens an international free trade regime. From their own perspective, high domestic food costs and farm subsidies reduce the competitiveness of their exports on the one hand, but on the other such protectionism is deeply institutionalized in national political structure and ideological consciousness. This conflict between national and international interests characterizes a central feature of the struggle to redefine the world order (McMichael 1992a). How the two states are negotiating this tension is the subject of this chapter.

One issue concerns rice protectionism. While Japanese food imports have doubled over the past decade (*Japan Economic Almanac* 1992:130)—building on a common import pattern for both Japan and South Korea of wheat, corn, soybeans, and animal protein—rice remains for both societies a symbol of national self-definition and food security. But in the context of the Uruguay Round, rice self-sufficiency threatens Japanese and South Korean negotiating power. In short, national policy implicates international relations, and vice versa.

After World War II nation building was both a domestic and a geopolitical project within the framework of the Cold War (Kaldor 1990). U.S. hegemonic designs inspired the reconstruction of the Japanese and South Korean states. Containment of East Asian communism dovetailed conveniently with the establishment of stable farm sectors, and these jointly underwrote industrialization with surplus farm products disposed of via U.S. food aid programs (Friedmann 1982). Central to this regional configuration was a particular feature of rice economies such as Japan and South Korea—an agricultural protection strongly biased toward food grains, explained by the "traditional lack of substitutability between rice and feed grains that made it possible to increase the price support on rice and, at the same time, to import feed grains without protection" (Hayami 1988:8). Such bifurcation between national production of food and the international supply of feed grains does not exist in Europe, where the food grains wheat and rye are substitutable for feed grains.

We argue that rice symbolizes the East Asian national farm sector and that its predicted liberalization would further internationalize these East Asian agrarian systems. By the same token, we suggest that precisely because of the availability of U.S. feed, and indeed, food, grains, alongside other imported fresh and processed foods, to supply new urban diets, rice protectionism could play its own idiosyncratic political and cultural role. This role includes supporting a complex of protectionist parastatal organizations and political constituencies that are currently eroding as the balance of political-economic forces shifts in each state.

As trade relations in the Pacific Rim have altered, at the expense of U.S. hegemony, the United States has resorted increasingly to bilateral pressure and the GATT to enforce market liberalization and gain support within these states for an agricultural reform process that would dismantle the national project. At the state level, the individual responses to these pressures reflect the different capacities of the two East Asian states in restructuring their farm sectors and reorienting their political institutions. Though there are many similarities between Japanese and South Korean agricultural systems, the relatively unspecialized nature of the South Korean farm sector renders it quite inflexible and more resistant to liberalization than the Japanese system. The variation in flexibility is rooted in historical differences between the two states, regarding the timing and ecology of their industrial strategies, which in turn express their mutual relationship, their regional position, and the structure of their respective political systems.

Substantively, then, we compare these two states as related parts of an international regime, governed by the rise and decline of U.S. hegemony. Neither the parts (states) nor the whole (international regime) can be understood without embedding a comparison of the two cases within a historical complex that situates each in time and space.[1] This

[1] The rationale for such a method of "incorporated comparison" that does not assume isolated and unrelated cases is spelled out in McMichael 1990 and 1992b. Here we should add that given our focus on domestic and international processes encouraging agricultural

historical complex is precisely the process of global structuring and restructuring of national and international relations as it unfolds in the regional complex centered on the United States, Japan, and South Korea.

National versus International Regulation

In general, the postwar Bretton Woods system of international political economy fostered national economic organization. Multilateral trade arrangements (institutionalized in the GATT and IMF support of stable exchanges, underwritten by the U.S. dollar) complemented national economies in which stabilization of farm and labor constituencies through protection of key economic sectors was legitimized in Keynesian policies of national regulation. Whether labeled "embedded liberalism" (Ruggie 1982) or "Fordist" political economy (Lipietz 1987), the ideal conditions of development came to be identified with the nation-state (Phillips 1977). Within this political environment, modeled upon Anglo-American experiences, agricultural modernization was considered to play a strategic role in the development process (Senghaas 1985:46–47). Indeed, in the state system under the *Pax Americana* a national farm sector became a universal ideal, where construction of national economies was modeled on the intersectoral relation between agriculture and industry identified with the U.S. economic trajectory (Friedmann and McMichael 1989). At the same time, U.S. postwar protection of the farm sector, institutionalized in GATT, encouraged replication of national farm protection in other nation-states (Friedmann 1993). Within this framework the U.S. model of food production and consumption was diffused as "the dominant national international model of regulation" (Tubiana 1989:25).

These East Asian agrarian systems are a case in point, where agro-industrialization, such as it was, followed the immediate postwar national goal of rural stability and rice self-sufficiency.[2] Nationally regu-

restructuring in relation to the nationally organized rice economy in each case, we have not examined the relationship between Japan and South Korea. Certainly the latter exports considerable amounts of fish and vegetables to Japan, but this exchange is insignificant compared with the relatively institutionalized "flying geese" or "product cycle" pattern characterizing industrial emulation of Japan by South Korea (and the other newly industrializing countries of Southeast Asia). Indeed, in 1972–76 Japanese investment in South Korea exceeded that of the United States by four times (Cumings 1984).

[2] In U.S. and Japanese agricultural development, according to Hayami and Ruttan (1985), differential land resources played an important role in inducing a particular pattern of techno-

lated rice and livestock sectors were complemented by imports of feed and food grains, supplying an intensive meat complex in addition to a general modernizing of diets. The resulting East Asian dependence on U.S. agricultural commodities served to intensify liberalization pressures on Japan and South Korea—indirectly by fueling (with wage foods) their export-manufacturing industries and directly through this regional division of labor. U.S. expectations of expanding commodity markets have increasingly been legitimized through neoclassical economic theory, as it gained the status of conventional economic wisdom in the 1980s.

The premise of neoclassical theory is that economic efficiency is equated with liberalization. It posits that liberalization follows logically from the growing redundancy of agricultural protection in the process of national economic modernization. Such protection is necessary during a period of adjustment of farming to industrialization, but it eventually constrains market efficiencies and the realization of comparative economic advantage (Hayami 1988:14–15). Based on the premise that "comparative advantage will continue to move away from agriculture" as an economy grows, the model explains agricultural protectionism as a policy adopted to avoid dependence on food imports (Anderson 1983:327).[3] Kym Anderson and Yujiro Hayami claimed: "This growth in agricultural protection is not confined to East Asia. A similar though somewhat more gradual trend has been occurring in most advanced industrial countries as their agricultural comparative advantages have declined" (1986:111).

Against this policy comparative advantage theorists argue that efficiency in the world food economy requires international forums such as the GATT "to help counter domestic political pressure for increases in agricultural protection" (Anderson 1983:328). National agricultural sectors are under pressure to reduce subsidies and import quotas that protect farm production in a universal process that, for metropolitan agricultures particularly, is understood as a process of "structural ad-

---

logical development—that is, mechanical (labor saving) versus biological and chemical (land saving), respectively. They observe that Japanese agriculture emphasized biological and chemical applications, such as breeding new varieties of rice and applying fertilizers, as early as the 1880s. It was only in the postwar period that significant progress in mechanization took place "in response to sharp increases in farm wage rates because of the rapid transfer of labor to the industrial and service sectors" (Hayami and Ruttan 1985:176). Japanese political conditions, however, have limited full rationalization of small-scale and part-time agriculture.

[3] The fact that this model is applied to Asian cases but ignores the U.S. experience (which appears to contradict it) highlights the reification common to economic modeling on the one hand and the selective understanding of economic history on the other.

justment" necessary to eliminate inefficiencies and secure a free trade regime (Hayami 1988:13). It is a prescription for deregulating national agricultural sectors, the implication being that food security is to be subordinated to, or accomplished by, elaboration of a global agro-food system based in comparative advantage.

In a broader sense, this challenge to nationalist regulatory structures is an attempt to dismantle the residual embedded liberalism of the postwar international economic order, as expressed in national social policies oriented to protecting specific economic sectors such as agriculture or subsectors such as rice or textiles (see Ruggie 1982). The new, prospective disembedded liberalism—as expressed in a GATT regime where trade discipline complemented the recent financial discipline of the IMF and the World Bank—would govern the world food economy at the expense of nationally organized farm sectors (McMichael 1992a, 1993b).

The salience of farm sectors in national political economy is both complex and specific to national trends and institutions. In general, Japanese agricultural modernization played a dynamic interactive role in industrialization (Hara 1990), whereas South Korea's exceptional export industrialism outstripped an agriculture governed more by political than economic forces (Moore 1986, Chung 1990). In addition, Japan's parliamentary system allows comparatively more decentralized policy negotiations among the central government, local government, and farmers' organizations. South Korea, by contrast, has a highly centralized executive–centered system, in which, until the late 1980s, local government and farm officials were appointed by the central government (Kim 1993).

Interaction among the political system, government policy, and industrial development pattern has greatly influenced the agricultural development paths of Japanese and South Korean agriculture, with three major differences. First, Japan has a high rate of part-time farming that involves high levels of off-farm employment (Jussaume 1991). It also has a related high rate of subcontracting *(keiretsu)*, in which a concern of many small firms is the cheaper labor in rural areas. By contrast, South Korean industrial subcontracting is currently concentrated in urban centers. Active government intervention, including designating export platforms and industrial zones in urban areas, was partially responsible for the concentration of industries in several big cities in South Korea (Moore 1985). As a consequence of this structural feature of Korean political economy, part-time farming in South Korea has been considerably less significant than that in Japan or Taiwan.

Second, farmers in Japan enjoy much greater income parity than farmers in South Korea. Per capita farm household income has exceeded per capita worker household income in Japan since the mid-1970s (Moore 1990:14), aided by the strong influence of the national agricultural cooperative Nokyo on the purchasing price of rice, and abundant rural off-farm employment opportunities. In South Korea, except for a brief period in the 1970s, the rural-urban income disparity has been a serious problem (Ban, Moon, and Perkins 1980; Burmeister 1992).

Third, the number of farm households in Japan has decreased more rapidly than that in South Korea. In South Korea, farm households decreased by only 3.5 percent between 1950 and 1980 (allthough the farm population itself fell by half between 1960 and 1980), whereas Japanese farm households decreased by 24.5 percent over the same period. Richard Moore (1990) attributes this difference to the two countries' inheritance patterns: in South Korea the eldest son inherits *most* of the land, whereas in Japan one heir usually inherits *all* of the land. Thus in South Korea land is more fragmented and thus there is a slower decline of numbers of farm households.

These differences between Japanese and South Korean agriculture might appear to result from a time lag, in which case we might expect South Korean replication of the Japanese pattern of farming rationalization in the future. However, it is more likely that there was, and will continue to be, some structural contrast between the two national political economies. Not only was South Korea's more centralized industrial and political system symbolic of a very late starter, but also it is likely that low-wage and low-tech subcontracting which supported off-farm work in Japan, is likely to move offshore under the current circumstances of East Asian regional integration and transnationalization of capital. This same movement now threatens Japanese farm household incomes (Moore 1991:24).

From a geopolitical perspective, industrial policy historically has been a priority of both national governments and of the postwar hegemonic project of the United States (Cumings 1984), and this policy was underwritten by U.S. agricultural exports. Rapid industrial expansion was strongly guided by state policy. Hagen Koo observes that "the South Korean economy is one of the capitalist world's most tightly supervised economies, with the government initiating almost every major investment by the private sector" (1987:172). As for Japan, Chalmers Johnson (1982) develops a detailed account of the crucial role played by the government bureaucracies in its industrial growth. If the roles played by Japanese and South Korean governments in industrial growth have been significant, the extent to which each government influenced the agricultural sector has been even greater (see Moore 1985, 1986 and Wade 1983 on South Korea; see ABARE 1988 and Riethmuller et al. 1988 on Japan). South Korean agriculture, closely regulated by the state via the National Agricultural Cooperative Federation, has been closely managed by the state, from provision of inputs and credit to pricing and disposal of the crop. Both governments distort agricultural prices (Amsden 1990) and protect domestic farmers by restricting agricultural imports, as well as by direct and considerable producer subsidies, especially in grain crops.

These protective measures now appear as residual national idiosyncracies, at least to proponents of liberalization. Food security is regarded as "a red herring in Japan's protectionist rhetoric" (Rothacher 1989:195), and for South Korea "food security is more an excuse for intervention than a real reason for it" (Anderson 1989:141). There may be truth to this observation from the perspective of existing capacities in the world food economy to supply East Asia (underlined by feed imports and growing high-value imports). The world economy, however, is complicated and secured by the existence of a state system that involves a variety of national political and cultural institutions that cannot be reduced so easily to economic rationality.

Advocacy of liberalization, however, concerns not only the issue of economic rationality. It fundamentally implicates the division of labor in the Pacific Rim and the rules of international political economy in a multipolar world as U.S. hegemony declines. That is, it has an ideological function, where, for example, U.S. demands for liberalization support general enhancement of its export markets.[4] In particular, U.S. demands play a surrogate role in the U.S. rivalry with the EC and Japan over the question of the regulation of a free trade regime (Akihiko 1991, McMichael 1993b).

Our argument, then, is that the struggle over liberalization expresses the broader rivalry over the emerging forms of global regulation. In the post–Bretton Woods era there has been an unstable movement away from national regulation and an erosion of national political economy

[4]The salience of farm liberalization in East Asia may be only symbolic, however, as "the impact of agricultural exports on the bilateral trade imbalances is minimal." In 1986, for instance, a study of the Congressional Research Service of the Library of Congress concluded that "the elimination of all Japanese agricultural import restraints could increase American sales by approximately $1 billion, but this would decrease the overall bilateral trade deficit by only 2 percent" (Higashi and Lauter 1987:126).

as international economic relations have intensified (Cox 1987, Kolko 1988, Held 1991). This trend has not gone unchallenged—we note especially the depth of resistance from farmer and environmental movements in East Asia—but it has been encouraged within states by a general redistribution of power from program-oriented ministries (e.g., social services, agriculture, education) to the central bank and to trade and finance ministeries (Canak 1989, Bienefeld 1990). This redistribution intensified during the debt regime via the enforcement of IMF conditionality, which promoted market liberalization, sectoral restructuring, privatization, and export promotion to service debt (Wood 1986). The next stage in the institutionalization of the mechanisms of global regulation would be completion of the Uruguay Round, rendering all states—some being more equal than others—subject to GATT discipline for unfair trade practices (Raghavan 1990:60). It is within this context that the issue of agricultural liberalization, and its institutional limits and possibilities in East Asian national political economy, is examined most fruitfully.

The Rise of the East Asian Agricultural Systems

In the early post–World War II period East Asia was involved in a dual process of rebuilding the national state and the nation-state system. Rebuilding the national state included rehabilitating the defunct social control systems of the government apparatus, bureaucracies, police, and revenue system. Rebuilding the nation-state system involved the incorporation of Japan and South Korea into the U.S.-led international network of commerce and economic and military aid. Both of these processes were conditioned by the Cold War context. The success of the Chinese Communist party and the outbreak of the Korean War were the two most important incidents governing U.S. policy of stabilizing and fostering East Asian capitalism.

Revitalizing the Japanese economy came to be considered important for the reorganization of world capitalism under the hegemony of the United States (Halliday, 1975). In the late 1940s the U.S. Occupation Force reversed its earlier opposition to the Zaibatsu (financal conglomerates) in Japan (Cumings 1984:17). Secretary of State John Foster Dulles initiated the "triangular program," linking the economies of the United States, Japan, and East and Southeast Asian states, whereby U.S. aid enabled Asian states to import Japanese products at the same time as Asian states obtained preferential access to the U.S. market (Wiley

1970). Integral to this program was the Mutual Security Act between the United States and Japan, part of which arranged for the financing of U.S. forces stationed in Japan with yen payments for imported U.S. agricultural surpluses (Kobayashi 1987:32).

Japan, and later South Korea, experienced rapid industrial growth fostered by U.S. aid, advice, and market access. Bruce Cumings has argued that these East Asian nations share the Bureaucratic-Authoritarian Industrializing Regimes (BAIR) model, characterized by "relative state autonomy, central coordination, bureaucratic short- and long-range planning, high flexibility in moving in and out of industrial sectors, private concentration in big conglomerates, exclusion of labor, exploitation of women, low expenditures on social welfare, . . . and authoritarian repression" (1987:81). Some characteristics of the BAIR model were supported by U.S. policies. These policies—notably military security, land reform, and food aid—conditioned the pattern of economic development, the relationship between the state and civil society, and the character of these East Asian agrarian systems.

Land reforms in Japan and South Korea were one response to the imminent social disorder after World War II and rising Communist popularity. In a context of antilandlordism, the U.S. Occupation Forces implemented land reforms that are still considered to be among the most successful in the capitalist world (Tuma 1965, Walinsky 1977, Pak 1982). The land reform destroyed tenancy and established a system of small-scale family farms in which the average farm was less than 1 hectare and there was a legal ceiling of 3 hectares, except for Hokkaido, Japan. The legal imits on land holdings, which lasted until 1962 in Japan and 1991 in South Korea, restricted the land market, precluding a process of differentiation of commercial farmers.[5] The unintended beneficiary of the land reform turned out to be the government in each case, because implementation of land reform revived previous bureaucratic systems. Although the U.S. Occupation Force was the major influence in the process, it had to rely on the traditional agrobureaucrats to conduct land survey and registration. As a move toward the democratization of Japanese society, the United States revived the agricultural cooperatives movement (Nokyo) in 1947. The new Nokyo, however,

[5] Although there was a significant time lag between Japan and South Korea with respect to legal limits on landholdings, it is important to note that the Food Control Act of 1942 in Japan established no-risk, price-subsidized rice growing, leaving farmers "little incentive either to increase the size of their farms or to diversify into other crops" (Akihiko 1991:3). This support system intensified in 1961–68, when rice prices rose rapidly, encouraging conversion from nonrice to rice crops (Moore 1991:26).

had democratic foundations (Moore 1990:141), although it operated as a patronage instrument of the Liberal Democratic party in organizing its rural constituency (Smith 1988:25). By contrast, the South Korean government completely dominated rural society, the agricultural cooperatives being parastatal arms to impose government policies.

Whereas land reform was crucial to reconstructing the national state in Japan and South Korea, U.S. food aid was a key element in the establishment of a regional nation-state system consistent with the U.S.-centered postwar world order. First, U.S. food aid provided the growing working classes of Japan and South Korea with cheap wage-foods (staple food commodities), sustaining a process of industrialization identified with success in this world order. Second, it contributed to a reshaping of East Asian diets, facilitating the long-term commercial incorporation of these East Asian agro-food systems into the world food regime (see Friedmann 1982, 1990). We will develop these points in the next section.

Through land reform and food aid the Japanese and South Korean states combined international and domestic forces in constructing their respective agrarian regimes. In the early years the configuration of the agricultural sector in each case was national, and "statist" to greater or lesser degrees, although modernization of diets was already under way in urban centers. The growing reliance on export industrialization in both cases intensified pressures to liberalize the East Asian agrarian systems.

Bifurcation of the East Asian Agro-Food Sectors

To grasp the essential dynamics of Japanese and South Korean agriculture, we employ the term *bifurcation* to describe selectively the establishment of complementary food supplies. Bifurcation of these East Asian agro-food sectors refers to their subdivision into a heavily protected national circuit of rice, as the basic food staple, and other agro-food circuits involving varying degrees of international commodity relations, such as the livestock complex and processed flour goods. Both states mediated the process of bifurcation through domestic and trade policies, but in different ways. The Japanese policy of agricultural rationalization (Agricultural Basic Law of 1961) separated subsidized rice farming from other subsectors that were exposed to increasing import liberalization—this policy dovetailed with the U.S. surplus food program (Kobayashi 1987:33, Ohno 1988). In the early 1970s the South

Korean government implemented a "protectionist agricultural price regime" in the rice subsector in response to decreasing U.S. concessionary grain sales (Burmeister 1992). Both the Japanese and the South Korean protection of rice farming were complemented dietarily by growing imports of U.S. farm products, especially feedstuffs.

Agro-food commodity imports were managed either within the framework of foreign aid or by state traders such as Japan's Food Agency or both. (Japan was more reliant on state trading as a means of agricultural protection) (ABARE 1988:36). Within this tightly regulated system, U.S. food aid to East Asia was particularly strategic in subsidizing industrialization and in the respective industrial export offensives of both states. There was a similar set of relationships—of food aid and a state-supported green revolution—in other Asian countries, such as India, Pakistan, and Indonesia, but without the close, state-coordinated industrial strategy, including management of the farm sector and its supplies of food and labor to the urban sector.

Cheap labor regimes depended on the government to maintain low food prices for the urban working class. Prices were kept low by the import of cheap U.S. grains under P.L. 480, largely on concessionary terms until the late 1960s. Food aid helped to suppress the prices of domestically produced grains, and the suppressed prices, in turn, undercut many farmers. Mick Moore summarized this mechanism: "Low grain prices kept labor costs down in two ways: directly, they reduced the reproduction costs and thus wage levels for the industrial labor force; and indirectly, they exerted a continual downward market pressure on urban wage rates by providing a continual supply of ex-rural job seekers eager to escape the even poorer material conditions of farming" (1986:105).

From 1957 to 1982 in South Korea more than twelve million rural people migrated to the cities, mostly to Seoul and Pusan (Chung 1990:143). The story for Japan was basically the same in its early years of postwar economic revival. While the Japanese government tried to increase rice production, cheap imported wheat lowered the reproduction (wage) costs of the working class. The result was a very competitive price of Japanese textiles and electronics in the world market (Hara 1990).

Consolidation of an urban proletariat in Japan and South Korea encouraged growing commercial imports from the United States. The consumption of wheat, never a popular basic grain among East Asian people, increased rapidly as a wage-food. Korean domestic wheat production fell by 86 percent between 1966 and 1977, as wheat imports rose by a factor of four (Wessel 1983:173). Animal protein consumption also grew apace. For Japan, food and feed imports increased from 3.6 to 18.6 metric tons during 1960–73 (Hemmi 1987:27), while during 1960–87 imports of major feed grains (corn and sorghum) increased fifteenfold, soybeans more than fourfold, and wheat twofold (Ohno 1988:20). Even when the concessionary terms for P.L. 480 ended, each country imported U.S. grains on commercial terms. As often celebrated by U.S. Department of Agriculture (USDA) officials, Japan and South Korea are exemplary cases of graduation from being recipients of food aid to important commercial buyers of U.S. food grains. In 1989 Japan and South Korea were the top two importers of U.S. agricultural commodities, importing US$7.4 billion and $2.5 billion, respectively (USDA 1990). The USDA, by cooperating with organizations such as Western Wheat Associates, played an important role in promoting the consumption of U.S. wheat in South Korea (Wessel 1983:173). In addition, both Japanese and South Korean governments were active in campaigning for the nutritional benefits of wheat, aided by nutritional and food scientists. They also provided free lunch bread to school children, as future "voluntary" consumers of wheat (Shinohara 1964).

Rapid economic growth, accompanied by the ideal of modernization, reshaped Korean social diets. The consumption of meat and dairy products spread from the upper class to other classes as the economy grew, encouraging a parallel, state-sponsored,[6] expansion of the domestic livestock industry: between 1975 and 1988 "Korean consumption of meat and seafood increased by around 65 percent . . . while Japanese consumption rose by almost 40 percent" (Harris and Dickson 1989:297). In South Korea the number of dairy cattle increased from 23,624 head in 1970 to 463,330 head in 1987, a nearly twentyfold increase in less than twenty years. In the meantime, the number of dairy farms increased tenfold over the same period (South Korean Ministry of Agriculture and Fisheries 1988). Processes of concentration and commercialization were more radical in the broiler and pork sectors. The expansion of the South Korean livestock industry relied heavily on imported feed grains, mostly from the United States, imports of corn and soybean increasing roughly tenfold from 1970 to 1984 (Anderson 1987:117). Although most production of assorted feeds using imported feed grains was strictly controlled by the South Korean government and

[6] Burmeister observes that "an important substitution program in beef production was envisioned as the 1980s sequel to the 1970s rice self-sufficiency campaign to help reinvigorate the farm economy" (1991:19).

National Livestock Cooperatives Federation, some U.S.-based grain corporations (e.g., Cargill, Inc.) set up poultry and animal feed operations in South Korea (Wessel 1983:174).[7]

In Japan similar changes began in the 1960s. The consumption of broilers and pork rose rapidly, while that of beef grew at a slower rate. Broiler production increased forty-four fold between 1960 and 1975, and pork production increased tenfold over the same period (Longworth 1983:33). While the dairy sector became a highly commercial business, organized by modern feedlot operations supplying supermarkets and meat processors under contract (Rothacher 1989:145), the traditional (Wagyu breed) beef sector remained a comparatively small-scale, specialized, feed-intensive activity undertaken by former owners of draft animals. The expansion of the Japanese animal protein subsector resulted in the explosion of imported feeds organized increasingly by large *sogo shosha* (trading company) feed suppliers (Rothacher 1989:64), as well as firms such as Cargill. For example, Japan imported 16 million tons of corn in 1988, 90 percent of which came from the United States (JETRO 1989). Imported feedstuffs provided two-thirds of Japanese livestock's nutritional requirements (Coyle 1983:34).

These developments reflect the bifurcation of Japanese and South Korean agriculture, which has, in many ways, become the axis of the liberalization controversy. On the one hand, the statist rice subsector is represented politically by active farm constituencies (including urban relatives) in each state and by their allies in the cooperative system. On the other hand, the nonrice agro-food interests (including food processors and other agro-industries) depend on international commodity circuits. The dairy and beef industries are increasingly specialized, with fattening systems dependent on imported feeds. This dependence, politically managed to reduce the cost of farm subsidies and stabilize wages (Yoshikazu 1985:46–47, Reithmuller, Wallace, and Tie 1988:156), puts governments squarely in the middle of the controversy—appearing to contradict food security arguments while raising effective protection by subsidizing intensive meat protection (Hillman and Rothenberg 1988:50).

In Japan the ruling Liberal Democratic party's (LDP) policy was always influenced by Nokyo. Nokyo, with monopolistic financial powers protected by the state (such as collecting 95 percent of all rice produced), has been an active interest group promoting agricultural protec-

[7] According to Kwang Jeon Lee (1989), the share of private plants over the National Livestock Cooperatives Federation's plant has grown to reach about 76 percent of total feed production.

tion (Smith 1988: 24–25). In addition, the rural gerrymander has enabled the LDP to remain in power for such a long period (Longworth 1983:62). Agricultural cooperatives have strongly influenced government rice policies. In the Agricultural Basic Law of 1961, for example, the Japanese government attempted to reduce the income disparity between the rural and urban population through agricultural diversification and modernization. Paradoxically, the increase of rural income was achieved less by the diversification than by the raising of the government rice purchase price. Agricultural cooperatives were active in lobbying the government to raise the purchase price of rice every fall. Meanwhile, the high rice price policy was possible because "rice ceased to be the basic wage-food crucial for industrial development" (Hara 1990:133) by the 1960s, having been displaced by wheat. As a result, virtually every farmer, especially the part-timers, produced rice. The government was forced to determine the purchase price of rice not on market terms but on sociopolitical terms. Every year, the Japanese government took the inflation rate, cost of production, and even the wage increase of the urban workers into consideration in determining the purchase price of rice. This method led to the extraordinary hike (through 1986) of Japanese rice prices compared with the price in international market, a source of complaint for free trade economists.

In Japan the LDP nurtured its rural constituency. But in South Korea agricultural policy making was not influenced by the local producer cooperatives. The executive-centered political structure granted the president power and did not allow lobbying room to the local farmers. The agricultural cooperatives were completely dominated by the central government (although recently this hierarchy has become more flexible). There was nothing cooperative in agricultural cooperatives *(Nonghyup)* other than the name (Ban et al. 1980). Meanwhile, farmers were forced directly and indirectly to join the organization because of its monopolization of important resources, ranging from credit to fertilizers to pesticides.

The 1970s saw a major change in agricultural policy in South Korea. This change was a result of two factors. First, the United States scaled down its concessionary grain exports. Second, the near defeat of President Park Chung-Hee in the 1971 election by Dae-Jung Kim, from the traditional rice basket of southwestern South Korea, led Park's regime to reevaluate the rural discontent with the income disparity between the rural and urban populations. The new agricultural policies included the initiation of the *Saemaul Undong* (New Village Movement) and the green revolution, to fulfill the goal of rice self-sufficiency. Through the

Saemaul Undong, the South Korean government invested extensively in rural areas. In addition to remodeling farmers' homes, the government committed money for roads and irrigation systems. This activity went hand in hand with the slogan of mechanization and rationalization of farming. It presaged a green revolution culminating in the invention of the "miracle" rice breed *tongil* (Burmeister 1988), encouraged by government-guaranteed purchase at a generous price.

These new agricultural policies of the South Korean government resulted in a short period of rising rural income in the 1970s. More important, however, these policies reestablished state control over the rural population. Nonghyup, as the distributor of chemical fertilizers, pesticides, and, most important, credit and government loans, was crucial in controlling the rural population and in guaranteeing rural votes for the governing party. Temporarily, South Korean farmers in the mid-1970s actually achieved high income growth due to the rice purchasing program and state credits, both of which operated through the Nonghyup. In the 1980s, however, farm incomes fell as rice farming suffered a growing productivity gap with the rest of the economy and social diets shifted further away from rice, creating twin problems of economic disarticulation (Burmeister 1992) and a lack of legitimacy for a government preoccupied with industrial economy (Moore 1985:180).

The evolution of East Asian farm policy within the Pacific Rim division of labor has thus bifurcated East Asian agricultures, at the same time that differentiated agrarian structures in Japan and South Korea reflect different modalities in export-led industrialization. In Japan the most important development has been the rise of part-time farmers. Industrial expansion in rural areas, the profitability of maintaining farm land (because of rising land and rice prices), and the expense of acquiring land led to an agriculture dominated by part-time farmers (Jussaume 1991). More than 85 percent of Japanese farmers were part-timers in the early 1990s. Many rural people commuted to nearby cities or worked at rural businesses and industries, using agricultural machinery to cultivate the fields over the weekend: "Rice growers bought farm machineries not to increase the size of their operations but to make part-time farming possible" (Hemmi 1987:28). By contrast, the South Korean industrialization pattern limited the availability of off-farm opportunities. The large conglomerates concentrated factories in a few big cities, differing from the Japanese pattern and even more so from that of the more decentralized Taiwan (see Ho 1982, Moore 1985). Hence South Korean farmers, 60 percent of whom are full time, are almost totally dependent on rice income, that is to say, on government subsidy.

Internationalization Trends

Because of certain global and domestic conditions, each government has reconsidered its statist, or national, agricultural policies. The global conditions stem from the collapse of the postwar international food order in 1973 (Friedmann 1982). This collapse led first to a "food crisis"—stimulating a new Japanese Food Security Policy (Kobayashi 1987:33)—followed by an unregulated and destabilizing trade crisis, as oversubsidized Western agricultural producers have competed intensely for export markets (Hathaway 1987), with the United States substituting commercial for concessional food exports (Friedmann 1990). The 25 percent increase in world grain trade within one year (Hathaway 1987:12) inflated food prices and exacerbated agricultural regulation in the West. In the EC, shortages legitimated agricultural subsidies, and in the traditional grain exporters, such as Australia and the United States, high prices promoted expansion (Alexandratos 1988:41). World grain trade expanded during 1974–80, the U.S. share rising to 60 percent (Insel 1985:897). By the late 1970s Europe rivaled the United States in grain exports as the Common Agricultural Policy (CAP) stabilized domestic prices by subsidizing exports, thereby intensifying dumping, depressing world prices, and further internationalizing diets (Watkins 1991:43).

The principle vehicle of agro-export expansion since the early 1970s has been the internationalization of U.S. agriculture via, in particular, the global diffusion of the intensive meat complex (Berlan 1989:129). The internationalization of U.S. agriculture was a deliberate policy decision to expand exports, in response to U.S. hegemonic decline (see Revel and Riboud 1986:58, McMichael 1994). Dismantling its regulatory structure (credit and price supports, including the food aid program), the Nixon government shifted from a national agricultural commodity supply management (direct aid to farmers and production quotas) to a demand management policy. The latter, geared to commercial exports, provides export incentives to farmers. Internationalization is expressed in monocultural export cropping and rising land concentration—both of which structure American farming into a profound dependence on the world market. The logic of this restructuring, in a context of extreme and unstable agricultural trade competition, is that the United States now seeks to liberalize East Asian economies. In addition, given the decline of U.S. hegemony, it is also logical for the United States to mobilize international enforcement mechanisms, such as the GATT (backed up by neomercantilist trade powers of retaliation against "unfair trade" in the Super 301 legislation of the mid-1980s), in its attempt to maintain a competitive advantage in the world economy (McMichael 1992a).

At the same time, the relative importance of Japanese and South Korean production and trade in the world economy has increased. Since the late 1970s, for example, while American industry lost 6 percent of its share of the world market, Japan gained 15 percent (Attali 1991:42–43). As Japanese and South Korean shares of global trade increased rapidly, trade surpluses, especially with the United States, became significant in the 1980s. From 1986 to 1989 the Japanese trade surplus with the United States exceeded US$200 billion. South Korea recorded $32 billion of trade surplus with the United States over the same period (Sohn 1991). The efforts by the United States to reduce its trade deficit through trade liberalization culminated in threatening to designate some nations "unfair trading partners" under Super 301. With this threat the United States won significant concessions from both Japan and South Korea (Levine 1990:21).

In the context of these trade realignments and their international political implications, the perspectives of the policy makers and the business groups of Japan and South Korea have become fluid. The policy makers in the Ministry of International Trade and Industry (MITI), the Ministry of Finance, and private business groups represented by Keidanren (the Federation of Economic Organizations) in Japan and the Economic Planning Board and Junkyungryun, a counterpart of Keidanren, in South Korea have often expressed the need to open up the whole economy, including capital markets. In 1982 Keidanren linked Japanese food security to a more competitive farm sector, claiming: "This situation not only makes the people's lives more difficult but also hinders Japan's food industry from becoming as competitive as its foreign counterparts" (quoted in Rothacher 1989:118). In 1990 the head of MITI, Kabun Mato, declared that Japan's rice protectionism was the Achilles' heel limiting its role in the GATT negotiations (Akihiko 1991:11). These groups are more influential in government policy making, and their industrial bias frequently overshadows the interest of more protectionist groups in both countries.

It has become evident that protection of farmers is increasingly problematic, not only because of its cost implications for government and individual and corporate consumers, but also because of the negative attention such protection receives in international forums. As a direct response to complaints by the U.S. trade representatives, both governments made trade concessions in 1988 on various agricultural commodities, such as oranges, dairy goods, and even beef. The issue at stake, however, is more than just trade—it is the very structure of the domestic agricultural sector.

The major contradiction concerns the issue of rice. Traditionally, rice has been considered special, even sacred, among Japanese and South Korean people. In addition, the postwar agricultural policies of both countries increased the importance of rice for farmers' income, especially through government price subsidies. As the response by Japanese farmers to the exhibition of U.S. rice showed Japanese rice farmers, considerably more entrenched in the state than their South Korean counterparts, resisted the liberalization of rice.[8] On the other hand, business groups often express their discontent with rice protection. As *The Japan Times* reported, "Apparently anxious about mounting frustration in the United States over the rice issue, business groups have been increasingly vocal in their calls for an end to the import ban" (April 23, 1991). In an action probably not unrelated to the 1986 complaint by the U.S. Rice Millers' Association, the Japanese government in the same year lowered the purchase price of domestic rice for the first time in thirty years (Kobayashi 1987). This trend will undoubtedly continue.

The South Korean rice issue is similar. Domestically, the overproduction and reduced consumption of rice stemming from reforms led the South Korean government to reconsider its rice purchasing policy. It is reported that government stockpiles of rice reached 2.9 million metric tons by the end of October 1990, costing the government US$2.8 billion. Meanwhile, per capita consumption of rice has been declining consistently: from 131.4 kg in 1981 to 119.5 kg in 1990.[9] To promote the consumption of rice, the government has allowed rice to be used for various alcoholic beverages, banned since 1964.[10] One of the most ambitious attempts by the government to solve the "rice question" has been the effort to install the land set-aside system *(hugyung bosangje)* to take advantage of GATT regulation 11.2-c, which allows the government to restrict the import of a particular crop if it is actively involved

[8] "Rice Dispute Intensifies in Japan," *Wall Street Journal,* April 17, 1991.
[9] "Changes Give Food for Thought," *Economic Report,* August 1991:40–42.
[10] "The Rice Soju Will Appear on Market" (in Korean), *Chosun Ilbo,* April 5, 1990.

in reducing the production of the domestic crop.[11] This regulation allows the South Korean government to argue that rice should be treated as a nontrade concern commodity in multilateral negotiations. In addition, the South Korean government, which gave up the dual price policy in the late 1970s (Kim and Joo 1982), plans to give up the rice purchasing policy completely,[12] reversing the miracle rice economy and further disarticulating rice growers from national economic growth (Kim 1991).

In both countries these changes have been taking place in tandem with the rapid decline of rural population and of the importance of agriculture in the national economy. The number of farmers and the share of agriculture in the GDP have declined significantly. The Korean agricultural sector's share of GDP fell from 37 percent in 1960 to 14 percent in 1983 (Borthwick 1992:278); Japanese agriculture's share fell from 23 percent in 1955 to 3 percent in 1985 (ABARE 1988:65). Most Japanese farmers commute to nearby cities, working at other jobs.[13] *The Economist* reported: "Part-time farmers have started leaving the land in droves. Most of them are aged 60 or so, the usual retirement age in Japan. Within a decade, half of them will be dead from hard work and old age . . . leaving the 470,000 hard-core professional farmers with more land to lease" (June 8, 1991:35). This statement records the restructuring of Japanese agriculture by commodity. In 1987 two-thirds of Japanese farms depended primarily on rice, and 40 percent cultivated less than 0.5 hectares, accounting for the considerable age and female majority among regular household members (ABARE 1988:72, 76, 81). As this demographic trend plays itself out within the rice complex, full-time core farmers either specialize in the expanding scale of livestock production (ABARE 1988:81) or enlarge farm scale through leasing arrangements encouraged by the government's Land Improvement Project, geared to complementing rice with mixed cropping (Moore 1991:31–33). Since most farmers in Japan have other sources of income, they are likely to be less resistant over the longer run to agricultural trade liberalization than South Korean farmers.

In South Korea farmers still account for about 18 percent of the total population, and roughly two-thirds of farmers are considered full-

[11] "UR Salshijang Goebang Daebi Jonnyak" (Strategies in response to the UR rice market liberalization), *Donga Ilbo*, April 19, 1991.

[12] "The Possibility of Opening Rice Market" (in Korean), *Chosun Ilbo*, April 25, 1991.

[13] Adachi Ikutsune argues: "In the logic of household economy, to farm is better than to have nothing to do. Nonagricultural employment offers sufficient income for them to lead an average living. . . . The outside-employed sons can grow rice in the paddyfield if it is 1 hectare or smaller by working Saturdays and Sundays with the help of machinery" (1985:21).

timers. It is likely that these farmers will resist government policy changes. Given the state's commitment to export industrialization, however, and its effective means of social control, the prospects for a continuing ban on rice imports are not very bright, especially since the farming population, already aging rapidly, will decline to about 10 percent by the year 2000 (*Business Korea,* April 1989:57).[14] Nevertheless, it is possible that a highly selective process of "decommodification" of rice farming in particular may occur, justified by the nonsurplus and solely domestic orientation of South Korea's rice subsector (Burmeister 1993).

There appears then to be a growing inevitability in the retreat of the state from the rural economy and society in Japan and South Korea. Not only has the international order changed with the demise of the *Pax Americana,* but domestic political-economic restructuring has also eroded the foundations of the poswar agrarian systems.[15] Though trade liberalization directly undermines the historical state-rice complex, other economic and institutional forces contribute to the general transformation of East Asian agriculture.

One such force is the livestock industry, increasingly dominated by agribusiness. From 1950 to 1985 the Japanese livestock industry increased its share of agricultural output from 7 to 27 percent, the share of rice declining by about 30 percent (ABARE 1988:66). Although 50 percent of cattle-raising operations remain small (often as a sideline to rice farming), a state-sponsored rationalization of beef processing, begun in 1960, has complemented the increasingly specialized and large-scale feedlot fattening systems (ABARE 1988:181, 197).

South Korea, lagging by a decade or so, followed. In the 1970s, with indirect and direct governmental assistance, some larger beef and dairy producers developed large ranches in former forest areas in South Korea (Moore 1985:141). Raising beef cattle was more closely regulated in

[14] Declining farming population and increasing availability of farmland will bring about the establishment of farming companies. As a matter of fact, in interviews with South Korean farmers (conducted in July 1991) it was revealed that Nonghyup is encouraging the setting up of farming companies by providing generous credits to young farmers.

[15] Paradoxically, the retreat of the state is a convergent outcome deriving from quite dissimilar conditions. In Japan the dilution of the farm population and its loss of electoral power—the share of farmers among LDP supporters fell from 29 percent in 1965 to 13 percent in 1985 (Akihiko 1991:12), and the LDP was defeated in 1993—have allowed such conditions to emerge over time as the state has undergone political transformation with the shift in the Japanese class structure. In contrast, the South Korean state, still relatively centralized in a political and spatial sense, is more likely to withdraw as it addresses rising sociopolitical demands of urban dwellers and seeks to continue to mollify them via an expanding industrial export economy.

the 1980s as part of a plan for price stabilization, which in turn aided concentration—the substantial rise in compound feed consumption (1975–87) attesting to the specialization and internationalization of the beef industry (Harris and Dickson 1989:297–99, Lee, Hadwiger, and Lee 1990:431–32). Vertical integration and incorporation into large multipurpose industrial conglomerates gathered pace in the 1980s (Moore 1985:180, Rothacher 1989:148).[16] The dairy, broiler, and pork industries were also organized by large agro-food capitals, including foreign agribusiness. South Korean government encouragement of Cargill, Inc., to directly invest in local soybean oil production was telling. An official at the Ministry of Finance commented on the Cargill decision: "We sympathize with them [domestic producers], but the time has arrived when they should come out from under the government's wing and stand on their own feet."[17]

East Asian industrialization, with its social, economic, and demographic changes, is transforming the agricultural cooperatives in each country. Nokyo has been criticized by business groups such as Keidanren and the mass media as impeding the modernization of agriculture and compromising Japanese consumers' purchasing power by raising domestic food prices (Smith 1988), and its material basis has been seriously challenged. With the deregulation of financial markets and interest rates, the invasion of large transnational financial institutions into the rural areas has weakened Nokyo as the dominant banking system there. Tageshi Domon (1991) analyzed the difficulties facing Nokyo's financial activities. Many farmers are leaving Nokyo, seeking higher interest rates for their savings account. No longer does Nokyo enjoy the monopoly of rural economic resources. Considering the importance of Nokyo's financial activities in its overall organization, it is no exaggeration to say that Nokyo's collapse as a vehicle of the farmers has already begun (Rothacher 1989, Domon, 1991).

In South Korea we see a similar process at one level; however, there are also peculiarities. Agricultural cooperatives themselves have become more democratic than they were. Nonghyup officials are now being elected directly by the local farmers, providing greater grassroots

[16] In Japan, Nokyo (with support from the Ministry of Agriculture, Forestry, and Fisheries) has gained the majority control of wholesaling in beef, organizing meat-processing companies around specialized meat-processing centers "to prepare boxed beef and consumer packs for immediate sale in supermarkets" (Rothacher 1989:78). And signifying Fordist methods of organizing agro-food commodity chains initiated in the integrated U.S. broiler firms (see Kim and Curry 1993), chicken-processing facilities "are now almost entirely controlled by the *soga shosha* affiliated feed suppliers" (Rothacher 1989:79).

[17] "Cargill Draws Up Battle Plans for Market Debut," *Economic Report*, September 1989:37.

input. On the basis of support by a relatively large farming population, Noghyup voices a strong protectionist agricultural policy, and an active farmers' movement has arisen. Agricultural cooperatives are trying to use survey results reflecting urban empathy with hardworking farmers to manipulate public opinion and future agricultural policies. They also have been trying to appeal to the urban consumers by emphasizing the importance of family, as many urban workers still have their extended family in the rural areas, and by advertising a centuries-old idea of agriculture as the basis of domestic economy *(nongbonsasang)*, which directly supports the food self-sufficiency idea. These activities embarrass the South Korean government, which is inclined toward liberalization under pressure of the U.S. trade representatives.

Park Joo-Kil, the senior delegate of South Korea's Economic Planning Board (EPB) to the GATT negotiations, carefully expressed his opinion in an interview on April 23, 1991, in Seoul. He stated that South Korea, along with Japan, would be forced to import at least 3 to 5 percent of its domestic rice. Park concluded the interview by commenting on the importance of agricultural restructuring.[18] In a domestic political economy that grants more power to finance and trade-related ministries and the EPB in policy-making processes, this comment expresses the internationalization of South Korean political economy. Ever since former president Park Chung-Hee started five-year economic planning in the early 1960s, the EPB has been at the center of organizing the South Korean economy. At the end of the 1970s, when the South Korean economy was facing its most serious crisis since the 1960s, the EPB formulated new guidelines favoring liberalization (see Haggard and Moon 1990).

The South Korean government has pursued various measures to make trade liberalization policies more appealing to the agricultural producers. The idea of rural industrialization to increase off-farm income is regularly cited by the government. The ambitious plan of the South Korean government to expand part-time farming is not realistic, however. The traditional rice basket of southwest Korea lacks the industrial infrastructure to attract industries. Companies prefer to build plants in the Southeast Asian countries rather than in rural South Korea. Not only are labor costs in rural areas higher, but there are few available rural laborers left because most of the young people have already migrated to the cities. It is estimated that nearly 33 percent of the farm population are over fifty years old. Considering these facts, the government plan for rural industrialization is little more than lip service.

[18] "The Possibility of Opening Rice Market" (in Korean), *Chosun Ilbo*, April 25, 1991.

In addition to rural industrialization, the South Korean government is trying to increase the size of landholding, to make agriculture more efficient by redistributing land from absentee owners via the Cultivated Land Corporation (Lee 1989:70). The scale of economy has been discussed as the panacea for the international competitiveness of South Korean agriculture. Early in 1991 the South Korea Ministry of Agriculture and Fisheries recommended to President Roh Tae-Woo a plan to raise the 3-hectare limit of farmland to 5–10 hectares, to increase agricultural competitiveness.[19] Although this measure may revitalize farmland markets, speculation on increasing land prices hampers the expansion of production.

In Japan, given the relatively more plural cooperative structure, the role of the state in agricultural adjustment is less direct. While historically the LDP has acted as the party of farmers, and Nokyo has mediated this relationship effectively by unifying rice producers, the trend of Japanese agriculture (especially the rise of part-time farming) has eroded this singular institutional complex (Gordon 1990:952). In 1990 there were about half a million full-time farmers, three-quarters of a million Class A farmers (mostly full time), and over three million Class B part-time households, who derived most of their income from nonfarm activities (Jussaume 1991). Since most full-time and Class A part-time farmers have diversified out of rice since the 1970s (ABARE 1988:112–9), it was Class B part-timers who depended most on rice, and therefore on rice subsidies—even though on average rice constitute less than 10 percent of income (Gordon 1990:953).

In sum, the rice-growing complex has a decining institutional weight in Japanese political economy. Part-time farmers have less to lose, vis-à-vis rice growing, by liberalization. Japanese farm policies are no longer responsible for the stability of rural communities, since rural industrialization underwrites off-farm income (Webb and Coyle 1992:129–30). The phenomenon of "hollowing out" of Japanese manufacturing, as industries move offshore, is likely to affect rural stability more than rice liberalization will—rendering a policy that supports GATT more realistic for the Japanese government. Nokyo—already under criticism from business groups, such as Keidanren, and the mass media as responsible for raising domestic food prices and thereby compromising agricultural modernization (inflating manufacturing costs) and Japanese consumer purchasing power (Smith 1988)—faces a challenge to its mo-

[19] "To Raise the Farmland Limit to 5–10 hectares" (in Korean), *Chosun Ilbo,* January 26, 1991.

nopoly of rural economic resources. Meanwhile, the 1993 defeat of the LDP signaled the end of a regime that rested in part on well-organized opposition to liberalization.

Under these circumstances, Japanese rice liberalization can be understood only as a quid pro quo for GATT support, upon which the Japanese political economy depends in the long run. The associated consolidation of rice farming on a growing scale, sanctioned in the 1992 reform program of the Japanese government, complements the growing confidence in bioindustrialization as the solution to the domestic food question. With its financial power and technology, Japanese food production is becoming a highly advanced industrial activity (Yoshikazu 1985), building on an already considerable biotechnical base (Wilkinson 1989) and an exploding consumerism. In short, Japanese agricultural policies depend more on overall economic concerns than on those of Japan's historical rice-farmer constituency and its institutional supports.

Conclusions

By comparing these East Asian agrarian systems as related, but different, constructs within the postwar Pacific Rim context, we can situate their national political-economic dynamics within a changing global framework that is characterized by an unstable displacement of national by international regulation. The regulatory rhetoric deriving from neoclassical economic theory isolates national economies and presumes a universal market waiting to be delivered from the womb of the late starter. A comparative-historical approach, however, identifies the geopolitics that stand behind these agrarian systems. The growth strategies of each state emerged within, and in turn transformed, the U.S. hegemonic project, as East Asian industrialization exacerbated U.S. trade deficits, relegating the rice and farm sectors to economic (but not political) insignificance. At the same time, the export-orientation of U.S. agriculture that emerged as a posthegemonic project assumed greater significance as East Asian diets shifted and as the United States sought to offset its deficits. Within this conjuncture the political goals of the United States in preserving advantage for its well-subsidized farmers in an increasingly competitive and unruly world economy have come up against the cultural protectionism and relative political inflexibilities of East Asian farming, however contradictory the neoliberal rhetoric. This is no simple matter of letting the market genie out of the bag.

The logic of the situation, however, leans toward increasing liberalization. There is resistance to liberalization, particularly among South Korean farmers, even though the politicization of agricultural "adjustment" has fractured South Korean politics, leaving the farm vote "up for grabs" (Burmeister 1990:717–19). Still, it is clear that state managers in each case favor liberalization, although the means to it are politically sensitive at the domestic level. There has been considerable unraveling of the coalitions underpinning the agrarian systems in each state—in part a result of domestic social-demographic change, and in part a result of internationalization (including financial deregulation) in the nonrice sectors of each state (see also McMichael 1993a).

A long-term perspective reveals that there is a certain structural logic in liberalization, insofar as this East Asian developmental model was premised on the U.S. military umbrella and access to the U.S. (world) market for industrial exports (McMichael 1987). This logic is playing itself out as it becomes clear that, by virtue of its relatively extreme dependence on internationial trade, Japan loses most in a failed Uruguay Round (Rowley 1992:35). In addition, the greater dependence of Japanese farmers on nonfarm income gives them a greater stake in Japanese economic growth and therefore in the success of the Uruguay Round (Gordon 1990:954).

With the transnationalization of economies and states (as a general process with particular variants) in the post–Bretton Woods era, the "embedded liberalism" in sectoral protection has become less a matter of national social policy and more a matter of bilateral and multilateral negotiation as states negotiate a competitive role in the world economy (Ruggie 1982, Gilpin 1987:404). Politically, these East Asian states are compelled to pursue agricultural liberalization as leverage in international negotiations to secure foreign markets for their prodigious industrial export regimes. Japan, however, has the capacity to compensate for downgrading its farm sector by offshore investment in the production of food and raw material for the Japanese market: It is a small irony that some of this investment is currently occurring in the United States itself.

References

Akihiko, Komani. 1991. *The Burden of Symbolism: Rice, Politics, and Japan.* USJP Occasional Paper no. 91–05. Harvard University: Program on U.S.-Japan Relations.

Alexandratos, Nikos, ed. 1988. *World Agriculture: Toward 2000*. New York: New York University Press.

Amsden, Alice. 1990. "Third World Industrialization: 'Global Fordism' or a New Model?" *New Left Review* 182: 5–33.

Anderson, Kym. 1983. "Growth of Agricultural Protection in East Asia." *Food Policy* 8: 327–36.

——. 1989. "Korea: A Case of Agricultural Protection." In *Food Price Policy in Asia*, ed. Terry Sicular, pp. 109–53. Ithaca: Cornell University Press.

Anderson, Kym, and Yujiro Hayami. 1986. *The Political Economy of Agricultural Protectionism in East Asia*. Sydney: Allen & Unwin.

Attali, Jacques. 1991. *Millenium: Winners and Losers in the Coming World Order*. New York: Times Books.

Australian Bureau of Agricultural and Resource Economics (ABARE). 1988. *Japanese Agricultural Policies*. Canberra: Australian Government Publishing Service.

Ban, Sung Hwan, Wong Moon, and Dwight H. Perkins. 1980. *Rural Development*. Cambridge: Harvard University Press.

Berlan, Jean-Pierre. 1989. "Capital Accumulation, Transformation of Agriculture, and the Agricultural Crisis: A Long-Term Perspective." In *Instability and Change in the World Economy*, ed. Arthur MacEwan and William Tabb, pp. 205–24. New York: Monthly Review Press.

Bienefeld, Manfred, 1990. "The Lessons of History and the Developing World." *Monthly Review* 41(3): 9–41.

Borthwick, Mark. 1992. *Pacific Century. The Emergence of Modern Pacific Asia*. Boulder, Colo.: Westview Press.

Burmeister, Larry. 1988. *Research, Realpolitik, and Development in Korea*. Boulder, Colo.: Westview Press.

——. 1990. "South Korea's Rural Development Dilemma." *Asian Survey* 30(7): 711–23.

——. 1992. "Korean Minifarm Agriculture: From Articulation to Disarticulation." *Journal of Developing Areas* 26(2): 145–68.

——. 1993. "Korean Agriculture: Protection or Phase Out?" Unpublished manuscript.

Canak, William. 1989. "Debt, Austerity, and Latin America in the New International Division of Labor." In *Lost Promises: Debt, Austerity, and Development in Latin America*, ed. William Canak, pp. 9–27. Boulder, Colo.: Westview Press.

Central Union of Agricultural Cooperatives (CUAC). 1990. *Agricultural Cooperative Movement in Japan*. Tokyo: CUAC.

Cerny, Philip G. 1991. "The Limits of Deregulation: Transnational Interpenetration and Policy Change." *European Journal of Political Research* 19: 173–96.

Chang, Yunshik. 1989. "Peasants Go to Town: The Rise of Commercial Farming in Korea." *Human Organization* 48(3): 236–51.

Chung, Jae-Yong. 1989. "Suipgaebang, Pokbaljikuneui Nongchon" (Import liberalization, rural areas are about to break down). *Sindonga* (November): 422–39.

Chung, Young-Il. 1990. "The Agricultural Foundation for Korean Industrial Development." In *The Economic Development of Japan and Korea*, ed. Chung Lee and Ippei Yamazawa, pp. 137–49. New York: Praeger.

Cox, Robert. 1987. *Production, Power, and World Order: Social Forces in the Making of History*. New York: Columbia University Press.

Coyle, William. 1983. "Japan's Feed-Livestock Economy." *Foreign Agricultural Economic Report No. 177*. Washington, D.C.: USDA-ERS.

Cumings, Bruce. 1984. "The Origins and Development of the Northeast Asian Political Economy." *International Organization* 38(1): 1–40.

Domon, Tageshi. 1991. "Nokyo O-koku no Naibu Moukaiga Maji Mat-ta" (Nokyo kingdom's internal collapse has begun). *Chuo Koron* January: 266–79.

Friedmann, Harriet. 1982. "The Political Economy of Food: The Rise and Fall of the Postwar International Food Order." *American Journal of Sociology* 88S: 248–86.

——. 1990. "The Origins of Third World Dependence." In *The Food Question: Profits or Prople?* ed. Henry Bernstein, Ben Crow, Maureen Mackintosh, and Charlotte Martin, pp. 13–31. New York: Monthly Review Press.

——. 1993. "The Political Economy of Food: A Global Crisis." *New Left Review* 197: 29–57.

Friedmann, Harriet, and Philip McMichael. 1989. "Agriculture and the State System: The Rise and Fall of National Agricultures, 1870 to the Present." *Sociologia Ruralis* 29(2): 93–117.

George, Aurelia, and Yujiro Hayami. 1986. "The Politics of Agricultural Protection in Japan." In *The Political Economy of Agricultural Protection,* ed. Kym Anderson and Yujro Hayami, pp. 91–110. Sydney: Allen and Unwin.

Gilpin, Robert. 1987. *The Political Economy of International Relations*. Princeton: Princeton University Press.

Gordon, Peter. 1990. "Rice Policy of Japan's LDP." *Asian Survey* 30(10): 943–58.

Haggard, Stephan, and Chung-In Moon. 1990. "Institution and Economic Policy: Theory and a Korean Case." *World Politics* 42(2): 210–37.

Halliday, Jon. 1975. *A Political History of Japanese Capitalism*. New York: Pantheon.

Hara, Yonosuke. 1990. "Agricultural Development and Policy in Modern Japan." In *The Economic Development of Japan and Korea,* ed. Chung Lee and Ippei Yamazawa, pp. 123–36. New York: Praeger.

Harris, David, and Andrew Dickson. 1989. "Korea's Beef Market and Demand for Imported Beef." *Agriculture and Resources Quarterly* 1(3): 294–304.

Harvey, David. 1990. *The Condition of Postmodernity*. Oxford: Basil Blackwell.

Hathaway, Dale E. 1987. *Agriculture and the GATT: Rewriting the Rules*. Washington, D.C.: Institute for International Economics.

Hayami, Yujiro. 1988. *Japanese Agriculture under Siege. The Political Economy of Agricultural Policies*. New York: St. Martin's Press.

Hayami, Yujiro, and Vernon Ruttan. 1985. *Agricultural Development: An International Perspective*. Baltimore: Johns Hopkins University Press.

Held, David. 1991. "Democracy, the Nation-State, and the Global System." *Economy and Society* 20(2): 138–72.

Hemmi, Kenzo. 1987. "Agricultural Reform Efforts in Japan." In *Agricultural Reform Efforts in the United States and Japan,* ed. D. Gale Johnson, pp. 24–46. New York: New York University Press.

Higashi, Chikara, and Peter Lauter. 1987. *The Internationalization of the Japanese Economy*. Boston: Kluwer Academic Publishers.

Hillman, Jimmye S., and Robert A. Rothenberg. 1988. *Agricultural Trade and Protection in Japan*. London: Gower Publishing Company.

Ho, Samuel. 1982. "Economic Development and Rural Industry in South Korea and Taiwan." *World Development* 10(11): 973–90.
Huang, Sophia, and William T. Coyle. 1989. "Structural Change in East Asian Agriculture." *Pacific Rim: Agriculture and Trade Report*, pp. 41–46. Washington, D.C.: USDA–Economic Research Service.
Ikutsune, Adachi. 1985. "Japanese Agriculture: Fallacy of the Revitalization Argument." *Japanese Economic Studies* 13(3): 3–33.
Insel, Barbara. 1985. "A World Awash in Grain." *Foreign Affairs* Spring: 892–911.
Japan External Trade Organization (JETRO). 1989. *White Paper on International Trade: Japan*. Tokyo: JETRO.
Johnson, Chalmers. 1982. *MITI and the Japanese Miracle*. Stanford: Stanford University Press.
——. 1987. "Political Institutions and Economic Performance: The Government-Business Relationship in Japan, South Korea, and Taiwan." In *The Political Economy of the New Asian Industrialism*, ed. Frederick Deyo, pp. 136–64. Ithaca: Cornell University Press.
Jones, Leroy. 1980. *Government, Business, and Entrepreneurship in Economic Development*. Cambridge: Harvard University Press.
Jussaume, Raymond. 1991. *Japanese Part-Time Farming*. Ames: Iowa State University Press.
Kaldor, Mary. 1990. *The Imaginary War: Understanding the East-West Conflict*. Oxford: Basil Blackwell.
Kim, Chul-Kyoo. 1993. "Capitalist Development, the State, and the Restructuring of Rural Social Relations in South Korea." Ph.D. diss., Cornell University, Ithaca, N.Y.
Kim, Chul-Kyoo, and James Curry. 1993. "Fordism, Flexible Specialization, and Agro-Industrial Restructuring: The Case of the U.S. Broiler Industry." *Sociologia Ruralis* 33(1): 61–80.
Kim, Dong-Hi, and Yong-Jae Joo. 1982. *Food Situation and Policies in the Republic of Korea*. Paris: OECD Development Centre.
Kim, Heung-Joong. 1991. "The Rise and Fall of Miracle Rice." *Economic Report* February: 56–57.
Kim, Yong-Jin. 1981. "Dairy Industry in Korea." *Asian Economies* 36: 5–17.
Kobayashi, Chutaro. 1987. "The Hard Rain of American Grain: An Historical Overview of Japanese Agricultural Policy." *AMPO Japan-Asia Quarterly Review* 19(2): 29–37.
Kolko, Joyce. 1988. *Restructuring the World Economy*. New York: Pantheon.
Komiya, Ryutaro, and Motoshige Itoh. 1988. "Japan's International Trade and Trade Policy, 1955–1984." In *The Political Economy of Japan*, ed. Takashi Inoguch and Daniel I. Okimoto, pp. 173–224. Stanford: Stanford University Press.
Koo, Hagen. 1987. "The Interplay of State, Social Class, and World System in East Asian Development." In *The Political Economy of the New Asian Industrialism*, ed. Frederic Deyo. Ithaca: Cornell University Press.
Kuznets, Paul. 1977. *Economic Growth and Structure in the Republic of Korea*. New Haven: Yale University Press.
Lee, Chung-Moo. 1989. "The Cause and Effects of Liberalization." *Economic Report* May: 68–70.

Lee, Kwang Jeon. 1989. "Dairy Industry in the Republic of Korea." *Asian Livestock* 14(2): 13–17, 14(3): 30–33.

Lee, Yong S., Don F. Hadwiger, and Chong-Bum Lee. 1990. "Agricultural Policy Making under International Pressures: The Case of South Korea, a Newly Industrialized Country." *Food Policy* 15(5): 418–33.

Levine, Paul. 1990. "U.S. Questions Korean Credibility on Market Opening." *Business Korea* December: 16–21.

Lipietz, Alain. 1987. *Mirages and Miracles: The Crisis of Global Fordism.* London: New Left Books.

Longworth, John. 1983. *Beef in Japan.* St. Lucia: University of Queensland Press.

McMichael, Philip. 1987. "Foundations of U.S./Japanese World-Economic Rivalry in the Pacific Rim." *Journal of Developing Societies* 3(1): 62–77.

——. 1990. "Incorporating Comparison within a World-Historical Perspective: An Alternative Comparative Method." *American Sociological Review* 55(2): 385–97.

——. 1992a. "National/International Tensions in the World Food Order: Contours of a New Food Regime." *Sociological Perspectives* 35(2): 343–65.

——. 1992b. "Rethinking Comparative Analysis in a Post-Developmentalist Context." *International Social Science Journal* 133: 351–65.

——. 1993a. "Agro-Food Restructuring in the Pacific Rim: A Comparative-International Perspective on Japan, South Korea, the United States, Australia, and Thailand." In *Pacific-Asia and the Future of the World-System,* ed. Ravi Arvind Palat, pp. 103–16. Westport, Conn.: Greenwood Press.

——. 1993b. "World Food System Restructuring under a GATT Regime." *Political Geography* 12(3): 198–214.

——. 1994. "GATT, Global Regulation, and the Construction of a New Hegemonic Order." In *Regulating Agriculture,* ed. Philip Lowe, Terry Marsden, and Sara Whatmore. London: David Fulton.

McMichael, Philip, and David Myhre. 1991. "Global Regulation vs. the Nation-State: Agro-Food Systems and the New Politics of Capital." *Capital and Class* 43: 83–105.

Moore, Mick. 1985. "Economic Growth and the Rise of Civil Society: Agricultures in Taiwan and South Korea." In *Developmental States in East Asia,* ed. Gordon White and Robert Wade, pp. 126–207. Brighton, U.K.: Institute for Development Studies.

——. 1986. "Mobilization and Disillusion in Rural South Korea." In *Food, the State, and International Political Economy,* ed. F. Lamond Tullis and W. Ladd Hollist, pp. 99–106. Lincoln: University of Nebraska Press.

Moore, Richard. 1990. *Japanese Agriculture: Patterns of Rural Development.* Boulder, Colo.: Westview Press.

——. 1991. "Strategies for Manipulating Japanese Rice Policy: Resistance and Compliance in Three Tohoku Villages." *Research in Economic Anthropology* 13: 19–65.

National Agricultural Cooperative Federation (NACF). 1979. *Urinaraui Saryosookpdonghyangkwa Dangmynkwaje* (Domestic supply and demand of feeds and ruminant problems). Seoul, Korea: Central Office of NACF.

Nishmimura, Koichi. 1970. *Agriculture in Japan.* Tokyo: Japan FAO Association.

Ohno, Kazuoki. 1987. "Nokyo: The 'Un'-Cooperative." *AMPO Japan-Asia Quarterly Review* 19(2): 25–28.

——. 1988. "Japanese Agriculture Today—Decaying at the Roots." *AMPO Japan-Asia Quarterly Review* 20(1, 2): 14–28.

Organization for Economic Co-operation and Development (OECD). 1987. *National Policies and Agricultural Trade: Japan*. Paris: OECD.

Pak, Ki-Hyuk. 1982. "Farmland Tenure in the Republic of Korea." In *Land Tenure and the Small Farmer in East Asia,* pp. 112–19. Taipei, R.O.C.: Food and Fertilizer Technology Center for the Asian and Pacific Region.

Phillips, Anne. 1977. "The Concept of 'Development.'" *Review of African Political Economy* 8: 7–20.

Raghavan, Chakravarthi. 1990. *Recolonization, GATT, the Uruguay Round and the Third World*. Penang: Third World Network.

Revel, Alain, and Christophe Riboud. 1986. *American Green Power*. Baltimore: Johns Hopkins University Press.

Riethmuller, Paul, Nancy Wallace, and Graewe Tie. 1988. "Government Intervention in Japanese Agriculture." *Quarterly Review of Rural Economy* 10(2): 154–63.

Rothacher, Albrecht. 1989. *Japan's Agro-Food-Sector: The Politics and Economics of Excess Protection*. New York: St. Martin's Press.

Rowley, Anthony. 1992. "Window of Opportunity." *Far Eastern Economic Review* 9 (January): 34–35.

Ruggie, John Gerard. 1982. "International Regimes, Transactions, and Change: Embedded Liberalism in the Postwar Economic Order." *International Organization* 36(2): 379–415.

Senghaas, Dieter. 1985. *The European Experience: A Historical Critique of Development Theory*. Leamington, U.K.: Berg Publishers.

Shinohara, Taizo. 1964. *Japanese Import Requirement: Projections of Agricultural Supply and Demand for 1965, 1970 and 1975*. Tokyo: Institute for Agricultural Economic Research, Department of Agricultural Economics, University of Tokyo.

Smith, Charles. 1988. "Monster in the Farmyard." *Far Eastern Economic Review* 17 (November): 24–25.

Sohn, Jie-Ae. 1991. "The Tug-of-War Heads toward Grand Finale." *Business Korea* January: 15–18.

South Korean Ministry of Agriculture and Fisheries. 1988. *Report*.

Tubiana, Laurence. 1989. "World Trade in Agricultural Products: From Global Regulation to Market Fragmentation." In *The International Farm Crisis,* ed. David Goodman and Michael Redclift, pp. 23–45. New York: St. Martin's Press.

Tuma, Elias. 1965. *Twenty-Six Centuries of Agrarian Reform*. Berkeley: University of California Press.

U.S. Department of Agriculture (USDA). 1989. *Pacific Rim. Agriculture and Trade Report*. Situation and Outlook Series. Washington, D.C.: USDA-ERS.

——. 1990. *Outlook for U.S. Agricultural Exports*. May. Washington, D.C.: USDA-ERS.

Wade, Robert. 1983. "South Korea's Agricultural Development: The Myth of Passive State." *Pacific Viewpoint* 24(1): 11–28.

Walinsky, Louis. 1977. *The Selected Papers of Wolf Ladejinsky: Agrarian Reform as Unfinished Business.* New York: Oxford University Press.

Wallerstein, Mitchel. 1980. *Food for War—Food for Peace.* Cambridge: MIT Press.

Watkins, Kevin. 1991. "Agriculture and Food Security in the GATT Uruguay Round." *Review of African Political Economy* 50: 38–50.

Webb, Alan J., and William T. Coyle. 1992. "Japanese Agriculture in the 1990s: An American Perspective." In *Agriculture and Trade in the Pacific,* ed. William T. Coyle, Dermot Hayes, and Hiroshi Yamauchi, pp. 119–32. Boulder, Colo.: Westview Press.

Wessel, James. 1983. *Trading the Future.* San Francisco: Institute for Food and Development Policy.

Wiley, Peter. 1970. "America's 'Pacific Rim' Strategy." *Australian Left Review* 26: 9–21.

Wilkinson, John. 1989. "The Reorganisation of the World Food System: Biotechnology and New Patterns of Demand." In *Prospects for the European Food System,* ed. Bruce Traill, pp. 99–121. London: Elsevier Applied Science.

Wood, Robert. 1986. *From Marshall Plan to Debt Crisis.* Berkeley: California University Press.

Yoshikazu, Kano. 1985. "Japanese Agriculture: It Can Be Revitalized." *Japanese Economic Studies* 13(3): 34–66.

Yoshioka, Yutaka. 1989. "Food Supply and the Japanese Concern over Food Security." *World Farmers' Times* 4(10): 6–9.

2

Sweden's 1990 Food Policy Reform: From Democratic Corporatism to Neoliberalism

David Vail

Sweden is not often mentioned in discussions of innovative agricultural policy. The notion of a distinctive "Swedish model" is usually evoked in reference to Sweden's extensive welfare state programs, its tradition of ultra-Keynesian macroeconomic policy, its active labor market measures, or its labor-management codetermination. Yet, after a decade of pronouncements about the urgent need for market-oriented agricultural policy reform by economists and politicians all around the industrial capitalist world, Sweden took the giant step from rhetoric to action, becoming, in June 1990,[1] the first highly protectionist nation to enact such a reform.

Although Sweden is a minor actor on the world agricultural stage, its food policy reform and the political economy behind it may be instructive. First, the new policy explicitly anticipates a resolution of the GATT's long, drawn-out Uruguay Round negotiations to reduce agricultural trade distortions and roll back the domestic policies that cause them. Second, other highly protectionist nations and the EC have also begun the process of deregulating their agricultural sectors, often along lines roughly analogous to the Swedish reform. Third, Sweden and several other small, affluent European nations with highly interventionist farm policies have applied for EC membership. They will have to accommodate their regimes to the Common Agricultural Policy (CAP); at

With thanks to my colleagues Knut Per Hasund and Lars Drake of the Swedish University of Agricultural Sciences, Uppsala.

[1] Two agricultural exporting nations, Australia and New Zealand, had already established market-oriented farm policies before the mid-1980s.

the same time, they will be in a position to influence the CAP via intra-EC alliances.

In a world where the "agrarian question" is taking on new shades of meaning, it is noteworthy that Sweden's Riksdag (parliament) enacted a *food* policy, not a new version of its sixty-year-old *agricultural* policy. The following are the core elements of the new food policy:

- A phased deregulation of domestic commodity markets over five years, eliminating farm price supports as the principal means of pursuing farm income and food security objectives
- Termination of export subsidies for surplus commodities produced at a cost above the international market price
- Preparation for the replacement of variable import levies by fixed and lower duties under a GATT agreement (variable levies insulated producers from international market signals and competition)
- Elimination of import protection for first-stage food-processing industries (largely owned by farmer cooperatives)
- Adoption of directly targeted measures to encourage production of public goods such as open and varied landscapes, wildlife habitat, and food security
- Transitional income compensation and farmland conversion supports

The immediate political-economic forces behind the reform can be interpreted in several ways. The cost of subsidizing the export of surplus products was no longer tenable because of the government's large budget deficits. The Social Democratic government sought to shore up its waning voter support by implementing a food policy to end double-digit price inflation. Readiness to abandon variable import levies was part of Sweden's strategy to promote nonagricultural interests in the Uruguay Round. Goodwill toward farmers was weakened by the public's growing awareness of industrial agriculture's negative effects on the environment and rural communities. Contraction in the number of farmers and the economic importance of agriculture had reached the point where the agro-industrial lobby could no longer sustain the political status quo.

In Sweden's multiparty political system, where a minority Social Democratic government had to engineer majority coalitions issue by issue, each of these factors undoubtedly had some influence. In sum, the momentum for policy reform in the late 1980s—like the policy inertia of the preceding half century—was overdetermined by a conjunction of multiple economic and noneconomic forces. This essay, however, attempts to explain the 1990 food policy at a more systemic level, as one moment in a fundamental restructuring of the core institutions and relationships that constitute Sweden's "negotiated economy." Sweden's distinction as a stable middle way, between unfettered capitalism and bureaucratic centralism, appears to be waning along with the twentieth century. A conundrum to grapple with is that, though Sweden has lagged somewhat behind the neoliberal trend that has swept the Western capitalist nations, it has taken the lead in deregulating the farm sector (Gunn 1989).[2]

The Negotiated Economy: A Working Definition

Klaus Nielsen and Ove Pedersen present a synopsis of the core institutions and relationships that constitute the negotiated economy. In advanced industrial capitalism, resource allocation and income distribution are interpreted as outcomes of the interplay between negotiation and three other core institutions—markets, democratic procedures (e.g., elections and parliaments), and administrative rules (i.e., bureaucracy). Mixed-economy models tend to emphasize only the latter three mechanisms. The forms and the relative influence of these four institutions are subject to change, and the ascendence of negotiation processes in twentieth-century Scandanavian social democracies is associated by Nielsen and Pedersen with their "homogeneous populations, relative symmetry of power in the labor-capital relationship, and long history of compromising, integrating and mediating" (1990:3). In other words, the negotiated economy is embedded in particular historically evolved class relations and sociocultural conditions.

As Nielsen and Pedersen stress, "The decisionmaking process is conducted via institutionalized negotiations between the relevant interested agents, who reach binding decisions typically based on discursive, political or moral imperatives rather than threats and economic incentives, even if such threats and rewards . . . might be essential elements of the framework around the negotiations" (1990:3). In the words of Albert Hirschman (1971), "voice," articulated through interest organizations, and "loyalty," developed through stable, long-term negotiated relationships, are central features, reinforced by the practical impossibility of "exiting" from those relationships. Competitive market pressures, elec-

[2] The term *neoliberalism* is used here as a shorthand for the espousal of competitive market solutions to socioeconomic problems and, by implication, the dismantling of much state regulation of economic activity. In the United States this creed might be labeled "neoconservatism."

toral politics, and regulatory structures can be viewed as imposing boundary conditions on negotiated relationships.

The state plays an integral role in negotiations between economic agents, and negotiation can be viewed as one dimension of complex democrat corporatist arrangements between the state and interest organizations. In a parliamentary democracy, corporatism puts the state in contradictory roles: it articulates and represents the public interest and also mediates conflicts among private interests. Indeed, Swedish corporatism antedates full democratization: the relatively autonomous nineteenth-century state administration invented several corporatist structures to carry out its dual roles (Rothstein 1988). In several contexts, including agricultural policy, a relatively autonomous state essentially created national organizations to represent hitherto unorganized economic interests in negotiations. Michele Micheletti (1990a) terms this strategy "sponsored pluralism."[3]

A central proposition of democratic corporatist theory is that neither the state nor interest organizations are fully independent actors: "The state delegates powers and participates in decisionmaking processes without full authority," while interest organizations are "integrated in the political process in a stable, long term manner" through a web of moral, cultural and discursive ties. A basic premise is that self-interest is "tamed," though not purged, through these socializing processes (Nielsen and Pedersen 1990:7, 13).

This behavioral interpretation contrasts sharply with public choice theory, the preferred model of political behavior employed by neoclassical economists. Public choice posits rational self-seeking (rent-seeking) behavior on the part of political actors and interprets negotiations as strategic games among such agents. (In the context of Swedish agricultural policy, this view is forcefully argued by Bolin, Meyerson, and Ståhl [1986] and Rabinowicz [1991]).

I accept "negotiated economy" as a useful description of the hybrid mode of production that took shape piecemeal in Sweden over a period of several decades after the ascent to power of the Social Democratic party (SAP) in 1932.[4] Specifically, the term *negotiation* describes one of the principal mechanisms used to constrain market forces and balance

[3] Private-interest organizations face "agency problems" analogous to those of the state in trying to aggregate their constituents' interests and mediate internal conflicts. Micheletti (1990b) presents an intriguing history of these tensions with the Federation of Swedish Farmers.

[4] *SAP* is the Swedish abbreviation for the Social Democratic party: Socialdemokratiska Arbetarpartiet.

conflicting economic interests in the Swedish agro-industrial system. The term also hints at nontrivial differences that distinguish Sweden's version of capitalism from the mixed economies of, say, Great Britain and the United States.

This analysis disagrees with Nielsen and Pedersen's claim that "the Scandanavian countries are increasingly assuming the character of a negotiated economy (1990:3)." To the contrary, recent evidence from agriculture and many other aspects of Swedish political economy supports the thesis that negotiation procedures are losing their hegemonic role, most importantly to market forces, but also to new bureaucratic rule-making procedures and, in the present period of fluid political party affiliations, to democratic processes. This trend, in fact, bears out another Nielsen and Pedersen proposition, namely, that the four core instruments governing the economy—negotiations, voting, administrative rules, and markets—"compete, disturb, and eventually supercede each other" (1990:3). A nonagricultural example illustrates this proposition: the intensification of international economic competition (markets) prompted an offensive by Swedish corporations to challenge both the fifty-year-old tradition of centralized wage bargaining (negotiation) and regulations governing paid employee sick leave. Externally induced change appears to be accelerating, and the final section of this essay briefly explores the likely impact on agricultural policy of a GATT agreement and Sweden's decision to seek membership in the EC.

Agriculture's Privileged Position in Advanced Capitalism

Swedish agricultural policy before 1990 was generically similar to that of several other nations and the EC, where the persistence of farm price supports, import protection, and export subsidies constitutes a major exception to contemporary capitalism's neoliberal tendency. Specific measures to insulate farm producers from domestic and international competition take several forms. The Swedish variant, like the EC's, centers on domestic farm price supports backed by variable import levies (indeed, Sweden pioneered this technique in the 1940s). The United States, in contrast, has maintained farm revenues via a combination of production loans repayable at target commodity prices and a deficiency payment mechanism.

Intensive farm input and growing farm output have been stimulated by a mix of artificially high prices and measures that encourage structural and technological rationalization of agriculture. These policies have been central causes of the overproduction tendency since the mid-1970s. In the face of well-organized political resistance to any fundamental policy reform, the EC, United States, and others have coped with their "butter mountains," "wine lakes," and "grain gluts" by adding costly new measures—export subsidies, acreage set-asides, "extensification" schemes, milk quotas, dairy herd buyouts—to their already complex policy regimes. In the 1980s these reactions contrasted sharply with the market-oriented deregulation of other economic sectors.

On a global scale, export dumping has frequently depressed international prices for food and feed grains below the break-even level for most farmers in most countries. Notwithstanding the existence of nearly a billion chronically malnourished people in the Third World, disruption of the former Soviet Union's food system, and ominous signs of deterioration in agro-ecological conditions, most agricultural economists predict that glut—not scarcity—will continue to plague capitalist agriculture for the remainder of this century (Berlan 1990, Brown 1992, Goodman and Redclift 1989, Ruttan 1990).[5]

The agricultural crisis, with its tangible symptoms of chronic surpluses, farm financial distress, environmental pollution, and rural socio-economic decay, was in the news all across Western Europe, North America, and Oceania in the 1980s. National debates over what to do about farm policy occurred in a setting of fiscal disarray, broad economic deregulation, and a swing toward neoliberal ideology. At the international level, the United States and the Cairns Group of fourteen agricultural exporters made reduction of agricultural trade distortions, and the domestic interventions that caused them, a top priority goal in the GATT negotiations. The Uruguay Round began in 1986, but was repeatedly set back by agricultural disputes, primarily between the EC and the United States. In late 1992 the last major hurdle, over EC oil seed subsidies, seemed to have been surmounted.

In the latter 1980s most nations and the EC temporized, reacting piecemeal to their food gluts and farm budget burdens, while politicians postured rhetorically about the need for a more thorough policy overhaul. Sweden became the first highly protectionist nation to enact such an overhaul, in June 1990. Swedish farm policy—unlike its innovative macroeconomic, social welfare, and labor market measures—has not

[5] The more localized environmental threats to long-term agricultural sustainability center on depletion of soil and groundwater, agricultural pollutants, loss of genetic diversity, and declining effectiveness of biocides. At an international and global level, acidification, stratospheric ozone depletion, and anthropogenic climate change are the most widely discussed threats (see Ruttan 1990, Brown 1992, Crosson 1992).

been widely studied outside Scandinavia. Nonetheless, the 1990 food policy should be instructive to agriculturalists since it presaged analogous (if less thorough) adjustments in other highly protectionist nations and modification of the EC's CAP in May 1992. Sweden's market-oriented reform should also interest social analysts as a signpost on the revisionist path taken by Sweden's Social Democratic party after almost sixty years of building a negotiated economy.

Democratic Corporatist Arrangements in the Swedish Model

Sweden's postwar social structure for capital accumulation and economic expansion was grounded in a classic labor-capital accord. It is commonly traced to the 1938 Saltsjöbaden Agreement, which legitimated centralized wage negotiations between the Swedish Employers' Federation and the major trade union federations. The state mediated labor-capital conflicts and articulated the public interest in wage restraint and minimization of productivity losses due to strikes and lockouts. The Social Democratic government's legitimacy in this corporatist arrangement was cemented by its commitment to maintaining full employment and avoiding intrusion into corporate management and investment decisions. The accord was given weight by the fact that the labor federations represented nearly all wage earners and gave the SAP its core voter support. One indication of the success of centralized wage negotiations in restraining the market's invisible hand was "wage solidarity": the practice of equalizing the compensation of workers in similar occupations, regardless of sectoral differences in productivity and profitability (Esping-Andersen 1985, Lundberg 1985, Milner 1989).

In the 1960s the trade union and SAP intellectuals breathed new life into the venerable democratic socialist vision of an economic democracy that would transcend full employment, universal entitlements, and equitable distribution of income. A series of laws mandating workplace codetermination was enacted in the 1970s, adding further layers of negotiation and consultation to the corporatist labor-capital relation. In 1984 legislation promoted a gradual socialization of capital ownership through Wage Earner Funds governed by representative boards.

Democratic corporatist institutions and negotiation processes have governed many collective actions beyond the core labor-capital relation. In Micheletti's terms, Sweden is "organizationally saturated": "encompassing interest organizations" serve as agents for groups as diverse as housing tenants, forest owners, parents of disabled children, performing artists, and of course farmers. Through propaganda and practice, negotiation of conflicts became a core value of Swedish political culture: the great majority of citizens approve the formal participation of interest organizations in design and implementation of policy, since they perceive the result to be an improvement in social harmony and the quality of life (Micheletti 1990a).

Interest organizations are typically represented on the state commissions that develop legislation affecting their members' interests. In several realms, tripartite negotiations substitute for free markets and individual contracts, as when a local housing authority meets with agents of landlords and tenants to set rent ceilings and guidelines for other rights and responsibilities. Interest organizations also have formal and informal roles in ground-level policy implementation. Bo Rothstein (1988) describes the latter as "administrative corporatism." He stresses the relatively minor role of government ministries vis-à-vis boards and agencies in policy execution.[6] Boards and agencies along with their various subcommittees and advisory councils, have substantial (often majority) interest-group representation. Finally, democratic corporatism's discursive dimension involves dense communication networks of conferences, interest-group publications, and mass media reportage.

Advocates of democratic corporatist arrangements contend that socialized and state-mediated economic decision making counteracts both the anarchy and the inequities of the market. It instills an ethic of social responsibility in participants that tames their pursuit of self-interest, it facilitates more farsighted decision making, and it results in socially superior outcomes (see Milner 1989, Pestoff 1989, Micheletti 1990b). Critics, particularly the public choice economists, argue that corporatist procedures foster the illusion that the common good prevails, when in reality private interests capture state institutions and use negotiations to win and protect economic rents (Bolin, Meyerson, and Ståhl 1986; Rabinowicz 1991). This controversy is explored further in the case of agricultural policy and in the context of international capitalist restructuring.

The Corporatist Policy Regime in Agriculture

The Social Democrats came to power in 1932 by forming a coalition with the Agrarian party that continued off and on for a quarter century.

[6] The ministries' central responsibility is to formulate policies rather than implement them.

In the 1930s Swedish farmers were mired in a long and severe economic depression; they also represented over 30 percent of voters, giving them an electoral clout that could not be ignored. The two parties negotiated a "cow trade," whereby agrarian interests accepted the Social Democrats' antidepression fiscal measures while the SAP tolerated farm price supports and import protection to revive agriculture. A long line of Social Democratic economists, notably Gunnar Myrdal in the 1930s and Assar Lindbeck in the 1960s, have been vocal critics of any farm policy that permanently insulated the farm economy from market pressures and incentives or pursued farm income goals via high food prices.[7] In general, SAP leaders have understood that import protection and price supports cause static economic inefficiencies, retard the flow of labor to higher productivity employment, and are distributionally regressive (by raising retail food prices). In the postwar setting of rapid economic growth and steadily rising nonfarm incomes, these costs and equity conflicts could be tolerated because Agrarian party support was needed for the Social Democrats' continued governance and the success of their larger socioeconomic agenda[8] (Vail, Drake, and Hasund 1994: chaps. 4, 5).

Nearly sixty years of market intervention should not be viewed simply as political expediency, however. For one thing, Sweden's military and diplomatic neutrality were used to legitimate the goals of food self-sufficiency and maintenance of farm production in remote, geographically disadvantaged regions. Furthermore, late-industrializing Sweden had a strong tradition of worker-peasant solidarity. Wage earners, most of them a generation or less removed from the farm, empathized with yeoman family farmers. The policy of income parity between efficient family farms and skilled industrial workers, established in 1947, was consonant with the labor unions' commitment to "wage solidarity."[9] Citizens' willingness to pay for costly farm supports was cemented by cultural values. Regardless of socioeconomic class, urban Swedes had—and have—a "place in the heart" for their ancestral villages and for the rich and varied agrarian landscape. Farmers were seen as the stewards

[7] Lindbeck subsequently left the SAP and became one of its most ardent critics.

[8] After the SAP-Agrarian coalition broke down the mid-1950s, the SAP's tactical motive for tolerating an unappealing agricultural policy was to help deter an Agrarian alliance with the bourgeois parties. The steady contraction of farm numbers and rural voter support eventually led the Agrarians to take the name Center party and seek a broader constituency, while remaining a voice for farm interests.

[9] Because income parity tended to slow the exodus from farming, farm policy was in perpetual tension with the macroeconomic goal of transferring labor—and other resources—to more productive sectors.

of this heritage. In sum, a moral dimension in Swedish political culture undergirded farm supports, as it does in most industrial nations. And in material terms, steadily rising consumer incomes and steadily rising farm productivity combined to keep food price increases within politically tolerable bounds (Frykman and Löfgren 1979, Drake 1987).

The Swedish Federation of Farmers (LRF) is the encompassing interest organization that eventually came to represent nearly all farmers. Because LRF's member cooperatives own most of the capital in farm input and primary food-processing industries, the group also often represents agro-industry in its negotiations and lobbying activities. There is thus a direct link between farmers and a significant element of the trade union movement. Although the LRF is the dominant agro-industrial interest organization today, the 1930s farm policy was not the outgrowth of pressure from any centralized farm organization. On the contrary, it was centralized price negotiations under the auspices of the National Agricultural Marketing Board that effectively forced Sweden's dispersed farm organizations to consolidate.[10] The LRF's political prominence thus has its origins in an interventionist policy regime.

Once established, the LRF became an encompassing, highly structured, and by all accounts extremely effective interest organization. It heavily influenced the legislative bodies and executive agencies that design, interpret, and execute farm policy. Its members have been prominent in the Agrarian party (later renamed the Center party) and the rightist Moderate party, and are well represented in the Riksdag. Until 1970 farmers made up over three-fourths of the Riksdag's Permanent Agriculture Committee (the 1990 figure was 41 percent) (Micheletti 1990b:111). Indeed, a Center party sheep farmer served as Sweden's prime minister during a period of nonsocialist coalition governments, from 1976 to 1982.

LRF has been formally represented on the commissions that design agricultural legislation (along with consumer and labor delegates). Its officers and technical experts sit on various advisory commissions and are routinely consulted by administrative agencies and the agricultural research establishment. As is common in other nations, the agricultural bureaucracy is largely staffed with people from farm backgrounds who retain personal ties to LRF members. At the local level, farmers have dominated most of the County Agricultural Boards. The LRF's influence within these institutions has persisted despite the steady contrac-

[10] Two distinct national organizations, a farmers' union and a federation of cooperatives, coexisted until the government induced them to merge and become the LRF in 1973.

tion of agriculture as a share of economic activity and farm people as a share of the electorate. (Less than 3 percent of the economically active population are now in agriculture.)

Public choice critics view this web of state and private interactions as evidence that the LRF has not been tamed by its multiform involvements in the policy apparatus. Rather, the nexus of the LRF, the Permanent Agriculture Committee, and the farm bureaucracy is viewed as a classic "iron triangle" that determines the policy agenda and subverts the public interest (Bolin, Meyerson, and Ståhl 1986; Rabinowicz 1991).

Semiannual negotiation of farm commodity prices (and some processed food prices) has been a site of direct encounter between agents for farmers and the Consumer Delegation. This relationship was another product of Social Democratic social engineering, since the Consumer Delegation was a legislative creation in 1963. Observers tend to agree that for most of the postwar period, producer interests prevailed over consumer interests in price setting. Indeed, for much of the period substantive bargaining was preempted by a legislated income parity formula that automatically passed productivity changes and higher input costs through to farmgate prices. The LRF's sophisticated media campaigns, emphasizing emotional issues such as national food security and preservation of the "living countryside," also helped maintain a climate of public opinion favorable to farmers—even in the mid-1980s when the cost of farm supports, via taxes and price inflation, reached nearly 2 percent of consumers' income (Micheletti 1990b, Rabinowicz 1991).

Economic Crisis and the Erosion of Social Democratic Governance

Swedish agricultural policy is complexly related to an unfolding crisis in the larger economy and to changes in the economic policy regime that have been precipitated by the crisis. These preconditions of the 1990 food policy reform can be presented in only a synoptic and stylized form here. (See Rothstein 1992 and Vail, Drake, and Hasund 1994 for a more thorough discussion.)

Sweden's long wave of stable economic expansion ended in the mid-1970s, as external pressures mounted and the class accord undergirding the postwar expansion began to unravel. Sweden's open economy was caught up in a global industrial and financial restructuring. Heavily dependent on energy imports and industrial exports, Sweden was particularly vulnerable to the two oil shocks and to the challenge by Japan and newly industrialized countries to its traditional heavy industry exports. Internal contradictions were amplified by the combination of imported stagflation and a secular productivity slowdown. The Social Democratic government's attempt to maintain full employment by expansionary macroeconomic policies intensified a serious wage drift above centrally negotiated norms. The breakdown of collective wage restraint in turn reinforced four other phenomena: the loss of international competitiveness, a corporate profit squeeze, capital flight, and the cost disease of the public sector.[11]

The mix of economic stagnation and expansionary fiscal policy precipitated a severe budget crisis: when unemployment reached the unprecedented level of 4 percent in the early 1980s, the budget deficit hit 12 percent of GDP. The private profit squeeze was a joint result of wage pressure and intense foreign competition, lagging productivity, and rising energy costs. The contraction in domestic capital accumulation had ominous implications for long-term employment and productivity growth (Lundberg 1985, Bosworth and Rivlin 1987, Svenska Handelsbanken 1990).

The twists and turns of Swedish politics since 1973 have been heavily influenced by the unfolding and still unresolved economic crisis. After more than four decades in power, the Social Democrats were voted out in 1976. But when economic conditions went from bad to worse under bourgeois coalition governments, the voters returned a minority SAP government in 1982. This status was perpetuated by the 1985 and 1988 election results. The economy, meanwhile, made a brief surge in 1983–85 and then settled into a unique full-employment stagnation, with GDP growth averaging 2 percent per year from 1986 to 1989, falling to 1 percent in 1990, and turning negative in the 1991 election year. Unemployment held below 2 percent for several years, but the resulting wage pressure (with sluggish productivity growth) pushed inflation to double digits in 1990 and brought growing recognition that higher unemployment was inevitable.[12]

The Social Democrats' internal splits, both strategic and ideological, were brought to a head by the lack of tools in the traditional policy kit to fix simultaneous economic stagnation, wage-push inflation, and corporate blackmail (i.e., the threat to move capital abroad). In the latter 1980s the SAP's neoliberal faction was strengthened by the chronic

[11] Swedish corporations have invested abroad for decades, but in the 1980s direct foreign investment increased almost sevenfold; the corresponding inflow of foreign capital grew only slowly and covered less than 20 percent of the outflow (Svenska Handelsbanken 1990:14).

[12] Unemployment crept upward in 1991, surged above 5 percent in 1992, and continued to rise in 1993.

low-level economic malaise, by the need to compromise with the (centrist) Liberal party to enact economic legislation, by a rightward drift in voter sentiment, and by the public's anxieties about Sweden's relationship to the EC.[13] Several years of sophisticated and increasingly aggressive free-market propaganda by the rightist Moderate party and the Swedish Employers Federation furthered the erosion of public confidence in the old corporatist remedies of negotiation, compromise, and bureaucratic intervention.

The Social Democratic response was revealed in a series of policy shifts:

- A pronounced shift from centralized negotiations and regulations toward market-oriented policies, particularly via financial market deregulation and the de facto end of centralized wage setting and wage solidarity
- Benign neglect of the wage earner funds through curtailment of dedicated tax revenues
- A supply-side tax reform that sharply reduced the top income tax rates and increased reliance on excise taxes (in practice, if not intent, the tax package is regressive)
- Creeping privatization of core public services, such as health and child care, to cope with the public sector's "sclerosis"
- A shift of responsibility for implementing the costly sick-pay system from the state to employers
- Abandonment of the commitment to (over)full employment; restrictive fiscal and monetary policies were expected to drive the unemployment rate above 3 percent by the end of 1991

These policies mark a detour from Sweden's middle way that is not likely to be either marginal or brief. Not only was the economic democracy project sidetracked, but policies that had long symbolized Sweden's commitment to social justice and economic equity were eroded. The failure of these maneuvers to capture the political center was revealed in the September 1991 election, when the Social Democrats' support fell to 38 percent; a four-party, center-right coalition took power with slightly less than a majority of Riksdag seats.

Internal Contradictions and External Forces Undermine Agricultural Corporatism

Sweden's agricultural policy crisis stemmed in part from the nonagricultural forces just described and in part from the food system's own

[13] After the 1988 election the SAP could no longer cobble together a Riksdag majority with its traditional ally, the Left (formerly Communist) party.

maturing contradictions. Three of these—related to economics, the environment, and national security—warrant a brief discussion.

By the late 1970s the stimulus of high farm prices and policy measures promoting technical and managerial "rationalization" had bred chronic and growing excess production. By 1984 wheat output was 65 percent above domestic demand; the figure was 27 percent each for milk and pork. Similar policies—with similar results—in the EC, the United States, and other nations flooded international markets and depressed prices. As a result of rising output and falling world prices, the cost of Swedish export subsidies quadrupled. At the same time, rising farm input costs, passed to consumers via the parity formula, pushed up retail food prices 66 percent between 1980 and 1984.[14] The media's attention to these issues, and growing criticism from the Social Democratic government, kindled popular skepticism of agricultural policy.

Popular disillusionment was reinforced by a spreading "green consciousness" that was fueled by mounting evidence of modern agriculture's adverse environmental impact, especially the degradation of aquatic ecosystems and the loss of open and varied landscape. By the time of the 1988 election, environmental concerns had become Swedish citizens' top political priority, and the Greens became the first new party to win seats in the Riksdag in seventy years.[15] Every other party was compelled to demonstrate its own "environmental friendliness." Out of a mix of genuine convictions and political expediency, economic interest organizations such as the LRF also scrambled for a place on the environmental bandwagon.

Farmers and agricultural policy were increasingly held accountable for the spread of monotonous monoculture landscapes, spruce plantations on arable land, nitrogen leaching into ground- and surface-waters, perceived chemical threats to food safety, and inhumane treatment of farm animals. From 1985 to 1990 a host of piecemeal "green" measures were enacted in response to these symptoms and demands (see Vail, Drake, and Hasund 1994: 6, 7).

As Cold War tensions waned in the latter 1980s, the conception of Swedish national security, including food security, was fundamentally questioned. Profound changes in the nature of warfare and in plausible scenarios of international crisis and blockade also cast doubt on the

[14] During this period budgetary pressure also led the government to phase out consumer subsidies on several food staples.

[15] Once in the parliament, the Green party did not impress voters with its legislative program or its participation in Sweden's brand of compromise and consensus politics. In 1991 the Greens' voter support fell below the 4 percent threshold needed for Riksdag representation.

relevance of peacetime self-sufficiency as a measure of preparedness. Of particular importance in new risk-benefit assessments was the fact that Swedish peacetime food production was heavily dependent on imported fuel and chemical inputs. Social Democratic leaders—and many others—could see no justification for a policy that induced excess production of foodstuffs at a cost far above world market prices, particularly an excess based on intensive use of inputs whose supply would presumably be disrupted in an international crisis.

Of course, a long-standing policy regime, popularly legitimated and backed by well-organized economic interests, is not destroyed overnight. However, the Social Democratic leaders set their sights on the archaic farm policy as an obvious target in their broader struggles to contain state expenditure and inflationary pressure, stimulate intersectoral labor mobility, preempt the bourgeois parties' calls for market-oriented deregulation, and project a green image to voters. For their own reasons, the bourgeois parties agreed with the SAP's means—agricultural policy reform—if not its political end of staying in power.

In the latter 1980s agricultural corporatism came under seige, as the LRF and the agricultural bureaucracy were outflanked on several fronts. First, legislation in 1985 terminated parity pricing, which had automatically passed higher input costs through to farm prices and had made producer-consumer negotiations a hollow exercise. Now the Consumer Delegation had greater scope to make its voice heard, and more than once it exercised the exit option by breaking off negotiations. Second, the government redefined national food security to emphasize larger emergency stockpiles of food and key inputs and deemphasize subsidized peacetime production. Third, the Environmental Protection Board and the Ministry of Environment and Energy, rather than agricultural agencies, were empowered to investigate agro-environmental problems and formulate policy responses. Fourth, when the government proposed a radical overhaul of farm policy in 1988, it shunned the corporatist process of a state commission with interest-group representation. Instead it formed a small working group of Riksdag members from the various parties with a staff of civil servants from nonagricultural backgrounds. To accentuate the SAP's priorities, the working group was charged to formulate a food policy, not another farm policy.

Much could be said about the likely effects of the food policy enacted in June 1990 with support from all but the Green party. The legislation reflects the parties' mastery of the art of compromise and is predictably generous to the thousands of farmers (especially milk and grain producers) who will suffer depressed incomes, decapitalization of assets, and pressure to leave agriculture. The new program is innovative in several ways; for example, it directly compensates farmers for providing public goods such as wildlife habitat and landscape amenities; it attempts to integrate part-time farming and value-added enterprises into regional development strategy; it privatizes most agricultural extension services; and it supports research and development of renewable agricultural bioenergy (see Vail, Drake, and Hasund 1994: chaps. 8, 9).

In this essay, the policy's core features are considered to be its economic liberalism and its attenuation of centralized corporatist decision making. After a five-year transition period, commodity markets will freely reflect domestic demand and supply forces. Price negotiations, central to the corporatist regime, and export subsidies have disappeared. Fixed import duties will replace variable levies, and the government has declared its willingness to cut duties deeply as part of a GATT resolution. (By implication, the political contest over farm prices has shifted from semiannual negotiations to the legislature's tariff-setting process.)

After a half century of privileged policy treatment, Swedish farm operators and agro-industrial workers are being edged out of secure corporatist niches toward the harsher world of market forces, but it would be wrong to infer that the negotiated economy is dead. Many elements of a new food policy regime are still being shaped, and a high-profile media debate and low-profile negotiations continue as various actors attempt to shape those elements. The LRF, with allies in the Center and Moderate parties, retains a prominent place in the contest over levels of import protection, in the design of new programs to promote conversion of redundant grain land to bioenergy crops and exportation of "clean and green" Swedish food products to the EC, and on county boards responsible for allocating most public funds for farmland preservation and rural development. As these examples suggest, agricultural reform in many respects entails a reconfiguration and decentralization of negotiated relationships, not their abolition.

Thousands of marginal farms, especially dairy operations in economically disadvantaged regions, are likely to be eliminated as prices fall toward world market levels. Despite targeted payments for landscape maintenance, as much as one-fifth of Sweden's 3 million arable hectares—especially land with high collective value in terms of biodiversity, scenic and recreational amenities, and cultural-historical significance—is likely to be afforested within ten to twenty years.[16]

[16] Unlike most western European nations, Sweden is largely a forested land. Large-scale afforestation, especially commercial spruce planting, thus has a net negative environmental and amenity effect (Drake 1987).

Taxpayers will pay nearly US$3 billion for temporary income compensation, land-conversion supports, and other measures. (This figure amounts to 2 percent of a year's GDP, spread over several years.) As consumers, however, Swedes will benefit from lower food prices, or at least slower inflation. Since farm-level revenues account for only about one-fourth of retail food costs, consumers' gains depend critically on measures to inject greater competition into the protected and highly concentrated food-processing and distribution sectors. Increased competition was a high priority for the SAP government, and a combination of antitrust and tariff reduction measures to break collusion and entry barriers in food distribution were enacted before the 1991 election.[17] The overall efficiency of resource allocation will be improved marginally by the flow of labor and capital from agriculture into higher productivity uses.

From a narrow economistic perspective, focusing on static allocative efficiency and material living standards, the market-oriented deregulation of Swedish agriculture could be called progress. If the new food policy, however, is viewed as one moment in the devolution of the larger social democratic project—the dismantling of a distinctive "Swedish model"—it is debatable whether the new policy regime represents social progress or regression, if environmental and cultural-historical values are given much weight. The Social Democrats' defeat in the September 1991 election and the current bourgeois coalition government's commitment to further economic liberalization suggest that the devolution will continue.[18]

Two likely changes in the international setting—a successful conclusion of the GATT negotiations and Swedish membership in the EC—would accelerate the present trends.

Swedish Agriculture in a "New World Order": GATT Agreement and EC Membership

The GATT: Dismantling Trade-Distorting Agricultural Policies

The Swedish food policy debate of the latter 1980s was premised on the assumption that a new GATT treaty would soon restrict the use of trade-distorting policies such as domestic price supports and export

[17] In 1992, for the first time in memory, retail food prices in Sweden actually declined.

[18] Two coalition members, the right-wing Moderates and the Center (formerly Agrarian) party, count on electoral support from farmers, but this has not significantly deflected the course of agricultural policy reform. In particular, the Moderates, who dominate the present coalition, are committed to a free market ideology.

subsidies. In pursuit of its most important objective, trade liberalization in the manufacturing and service industries, Sweden has been willing to make concessions on agriculture and has generally supported the Cairns Group's position.

At the end of 1992 the United States and the EC seemed at last to have resolved the conflicts that had obstructed agreement for six years. At that time, it appeared that the compromise solution would resemble the proposal made late in 1991 by Arthur Dunkel, director of the GATT: variable import levies would be replaced by fixed duties, export subsidies would be curtailed, income support would gradually be uncoupled from farm prices, and state budget support to agriculture would be cut by one-third or more by the end of the 1990s (Viatte and Cahill 1992). As of October 1993 no final agreement had been reached on these conditions.

But Sweden is already well on the way to meeting these conditions. Limited analysis of such a partial and gradual deregulation suggests that most of Swedish agriculture would probably remain competitive in home markets through the end of this century. Imports of cheese and some other animal products might increase significantly, however, and the long-term decline of farming in northern Sweden would accelerate if special regional price and income supports had to be abandoned. Much would depend on whether the EC and United States could dispose of surplus meat and dairy stockpiles without severely depressing international prices for several years, which could drive efficient producers in the southern Sweden plains out of business (Vail, Drake, and Hasund 1994: chap. 9).

As public concerns about environmental quality and food safety intensify in the industrial nations, agricultural trade controversies will inevitably persist in new forms after the Uruguay Round is concluded. Indeed, a "green round" was proposed, to come to grips with import restrictions justified by food safety, environmental protection, and animal-rights considerations (Uimonen 1992). Carol Kramer summed up the economic stakes that are bound up with green trade restrictions: "One of the major agricultural proposals being negotiated under the GATT is the harmonization of national food safety standards. The motivation for globally establishing and maintaining the same food safety standards is that divergent national standards have been used as nontariff barriers to trade. There is a fear that nations will increasingly resort to these standards if and when other trade barriers are eliminated" (1991:10).

Swedish agriculture could be profoundly affected by the outcome of such negotiations. Its taxes on fertilizers and biocides, strict limits on chemical residues in food, winter cover crop requirements, tough regulations on manure management, and exceptional livestock husbandry standards all increase production costs vis-à-vis foreign competitors. Sweden thus has a stake in creating a common negotiating front with other nations such as Denmark, Germany, Japan, and Switzerland, which have also set high green standards. To do so would probably require that Sweden modify the standards and instruments it introduced unilaterally at the end of the 1980s, but such a sacrifice of national sovereignty would probably be in Sweden's interest, since the outcome—more rigorous green standards in other advanced industrial nations—would improve Swedish economic competitiveness while also fulfilling its citizens' green convictions. (These issues are elaborated in Vail, Drake, and Hasund 1994: chaps. 9–11.)

Sweden in the European Community

Within a few months of the food policy legislation, the Riksdag voted nearly unanimously to apply for full EC membership. This move would certainly be Sweden's most profound political and economic redirection in the twentieth century. Sweden's state of political flux is reflected in the fact that two decisions with such momentous and potentially conflicting implications as the food policy reform and the EC application were made in the same year. One might well ask whether it was rational to launch a deregulation of the farm economy only to subject it to a new regulatory regime under the CAP within a few years. In any event, the newly created Swedish Board of Agriculture's motto conveys a distinctly internationalist vision: "We shall work for a competitive, environmentally friendly and livestock-friendly food production in a European perspective."

Negotiations began in early 1993, with membership possible by the end of 1994, pending the results of a popular referendum. All the major political parties officially support EC membership, but Swedish voters are deeply divided. The Danish rejection of the Maastricht Treaty for European Union in June 1992 at least temporarily strengthened adverse opinion. The timetable could be affected by delays in Maastricht's adoption and would certainly be affected by its defeat. Even its defeat, however, would not spell the end of the Policy or dissuade Sweden's coalition government from seeking membership.

For Swedish society and the economy as a whole, it is hard to avoid the conclusion that EC membership would decisively erode the relatively autonomous legal structures, policy framework, and administrative arrangements that have sustained the negotiated economy and allowed Sweden to follow a distinctive middle way for over a half century. For Swedish agriculture, however, the implications are less clear. A central question is what the CAP itself will look like in the future. After all, the CAP has come under intense and ceaseless criticism because of its spiraling budget cost, obvious economic deficiencies, and dubious distinction as the most serious roadblock to a GATT agreement. With the exception of the GATT dimension, the economic defects of the CAP and of Sweden's old agricultural policy are quite similar.

In 1991 EC Agricultural Commissioner Ray MacSharry proposed a far-reaching reform of the CAP, along lines loosely analogous to Sweden's 1990 food policy. In May 1992, after a year of intense debate—and strenuous protest from the EC's well-organized farm lobby—the EC's agricultural ministers adopted a watered-down version of the MacSharry Plan, which nonetheless promised the most far-reaching change in the CAP's history. Very briefly, the new policy calls for gradual cuts in grain, beef, and butter prices; lower milk quotas; a halving of public purchases of surplus meat; and acreage set-asides for large farms. In return, producers of price-regulated commodities will receive compensation for lost revenue on a per-hectare or per-animal basis. The compensation-cum-set-aside mechanism is sufficiently like the United States' deficiency payment system to suggest that a tactical objective was to break the GATT deadlock.[19]

In sum, the CAP will take a reformist path with some parallels to the one Sweden has followed since 1990. Critical questions for the future of Swedish agriculture once it comes under CAP's jurisdiction are what levels of price support and import protection the CAP will retain; what impact communitywide competition in agriculture and food processing will have on Swedish farmers; how Sweden's geographically disadvantaged regions will fare under the EC's structural measures and regional economic policy; and what the EC's green standards will be.

Swedish dairy farmers would presumably be protected by the CAP's milk quota system, and Sweden could even capitalize on its environmentally friendly reputation by capturing some export niche markets. However, cash grain and oil seed producers would have to compete with formidable French rivals, and livestock producers would go *mano a mano* against the intensive, industrialized production and processing systems of Denmark and the Netherlands. The biggest shock could come from the free entry of lower-cost processed and manufactured foods from the EC's agribusiness giants. The northern two-thirds of Sweden faces climatic conditions far harsher than designated "disad-

[19] "EC Farm Policy: Getting Better," *The Economist*, May 23, 1992:55.

vantaged regions" in, say, Britain or Italy. Unless Sweden and two other prospective EC members, Finland and Norway, can win special exemptions for their subarctic regions, only hobby farming is likely to survive in these areas beyond the turn of the century (see Vail, Drake, and Hasund 1994: chap. 9).

Perhaps Sweden's best hope is that its market-oriented agricultural policy and its wide array of green initiatives will make it an attractive ally, possibly even something of a model, for other northern EC members. The Danish foreign minister welcomed Sweden's 1990 application precisely on the grounds that Sweden would strengthen the intra-EC forces pressing for more progressive environmental initiatives. And the fact that three other economically and environmentally advanced nations (Austria, Finland, and Norway) are all currently negotiating membership also suggests that the EC's environmentally progressive alliance will be strengthened. The community is just beginning to shape a framework of common environmental standards, and its regional development policy is in flux, partly because of the 1992 CAP reform. In this setting, Sweden would benefit greatly if EC policies converged in the direction of its own (Tobey and Ervin 1990).

The "Agrarian Question" in a New Millennium

In general, Sweden has lagged behind other advanced capitalist nations in its neoliberal drift, the pace slowed and the direction modified by the institutional traditions and political culture of the negotiated economy. In agricultural policy, however, Sweden acted in advance of the other highly protectionist nations to increase the scope of market forces and reduce the influence of negotiations and the state administration. Although the pace and extent of Swedish reform may be exceptional, the trend toward greater market orientation in agricultural policy seems to be nearly universal. Austria, for example, has eliminated its food export subsidies; Norway and Switzerland are uncoupling farm income support from commodity pricing; and Japan cut its support price for rice four times between 1986 and 1991 (Rauch 1991). More important on a world scale is the fact that the United States' 1990 Farm Bill furthered the deregulation process begun in 1985, and the EC, as described above, will gradually and partially reduce the CAP's distortion of domestic and international markets.

The tendency toward convergent agricultural policies in advanced capitalism seems certain to gain momentum: on a European scale, as Sweden and other nations join or develop other forms of association with the EC, and on a global scale, as the Uruguay Round is brought to a conclusion.

The convergence process has been driven primarily by economic forces. But agriculture is taking on a new mix of meanings and priorities, especially in advanced industrial societies such as Sweden. The old agrarian politics, centering on food security, economic rationalization, and equity for family farmers, is giving way to a new politics. At its core are widespread public concerns about longer-term sustainability of production capacity, cultural-historical landscapes, rural communities, biological diversity, and aquatic ecosystems. The preceding discussion of rules governing international trade and the CAP's evolution hinted that harmonizing these noneconomic aspects of agricultural and rural development policies will be a major challenge during the remainder of this century. I would claim that Sweden is also in the vanguard in designing policies to foster a sustainable agriculture and that its green initiatives hold valuable lessons for other nations. Those claims, however, must be the subject of another essay.

References

Berlan, Jean-Pierre. 1990. "Capital Accumulation, Transformation of Agriculture, and the Agricultural Crisis." In *Instability and Change in the World Economy*, ed. Arthur MacEwan and William Tabb, pp. 205–24. New York: Monthly Review Press.

Bolin, Olof, Per-Martin Meyerson, and Ingemar Ståhl. 1986. *The Political Economy of the Food Sector.* Stockholm: SNS Förlag.

Bosworth, Barry, and Alice Rivlin, ed. 1987. *The Swedish Economy.* Washington, D.C.: Brookings Institution.

Brown, Lester. 1992. "World Grain Takes a Spill." *Worldwatch* 3: 35–36.

Crosson, Pierre. 1992. "Sustainable Agriculture." *Resources* no. 106: 14–17.

Drake, Lars. 1987. "The Value of Preserving the Agricultural Landscape." Paper presented at the Fifth European Congress of Agricultural Economists, Balatoaszeplak, Hungary.

Esping-Andersen, Gøsta. 1985. *Politics against Markets: The Social Democratic Road to Power.* Princeton: Princeton University Press.

Frykman, Jonas, and Orvar Löfgren. 1979. *Culture Builders.* New Brunswick: Rutgers University Press.

Goodman, David. 1991. "Some Recent Tendencies in the Industrial Reorganization of the Agri-Food System." In *Towards a New Political Economy of Agriculture*, ed. William Friedland et al., pp. 37–64. Boulder, Colo.: Westview Press.

Goodman, David, and Michael Redclift, eds. 1989. *The International Farm Crisis.* New York: St. Martin's Press.

Gunn, Simon. 1989. *Revolution on the Right.* London: Pluto Press.

Hirschman, Albert. 1971. *Exit, Voice and Loyalty.* Cambridge: Harvard University Press.
Kramer, Carol. 1991. "Implications of the Hormone Controversy for International Food Safety Standards." *Resources* no. 105: 12–14.
Lundberg, Erik. 1985. "The Rise and Fall of the Swedish Model." *Journal of Economic Literature* 23(1): 1–36.
Micheletti, Michele. 1990a. "Interest Groups in Post-Industrial Sweden." In *Jahrbuch zur Staats- und Verwaltungswissenschaft 1990.* Berlin: Max Planck Institute.
——. 1990b. *The Swedish Farmers' Movement and Government Agricultural Policy.* New York: Praeger.
Milner, Henry. 1989. *Sweden: Social Democracy in Practice.* New York: Oxford University Press.
Nielsen, Klaus, and Ove Pedersen. 1990. "From the Mixed Economy to the Negotiated Economy." Paper presented at the Society for Advancement of Socioeconomics Conference, Washington, D.C. (also forthcoming in *Morality, Rationality, and Efficiency: Perspectives on Socio-Economics,* ed. R. Coughlin. New York: M. E. Sharpe).
Pestoff, Victor, ed. 1989. *Organizations in Negotiated Economies.* Report no. 2. University of Stockholm, Department of Business Administration.
Pestoff, Victor. 1991. "The Demise of the Swedish Model and the Rise of Organized Business as a Major Political Actor." Paper presented at the Society for Advancement of Socioeconomics Conference, Stockholm, June.
Rabinowicz, Ewa. 1991. "Agricultural Policy: Old Wine in New Bottles." *Report 32, The Study of Power and Democracy in Sweden.* Uppsala: Maktutredning.
Rauch, Jonathan. 1991. "Spilled Milk." *National Journal* (September 14): 2210–14.
Rothstein, Bo. 1988. "Social Classes and Political Institutions: The Roots of Swedish Corporatism." *Report 24, The Study of Power and Democracy in Sweden.* Uppsala: Maktutredning.
——. 1992. "The Crisis of the Swedish Social Democrats and the Future of the Universal Welfare State." Paper presented at the Harvard University Center for European Studies, Cambridge, Mass.,
Ruttan, Raymond. 1990. "Scientific, Technical, Resource, Environmental, and Health Constraints on Sustainable Growth in Agricultural Production." Paper presented at the Symposium on Population-Environment Dynamics, University of Michigan, Ann Arbor, October.
Svenska Handelsbanken. 1990. *Sweden in the World Economy.* Stockholm.
Tobey, James, and David Ervin. 1990. "Environmental Considerations of the Single European Act." Paper presented at the USDA conference, Europe 1992: The Future for World Agriculture, Washington, D.C., November.
Uimonen, Peter. 1992. "Trade Policies and the Environment." *Finance and Development* 29(2): 26–27.
Vail, David, Lars Drake, and Knut Per Hasund. 1994. *The Greening of Agricultural Policy in Industrial Societies: Swedish Reforms in Comparative Perspective.* Ithaca: Cornell University Press.
Viatte, Gérard, and Carmel Cahill. 1991. "The Resistance to Agricultural Reform." *The OECD Observer* 171 (August).

3

Agricultural Change in the Semiperiphery: The Murray-Darling Basin, Australia

Geoffrey Lawrence and Frank Vanclay

Despite exhibiting the expenditure and lifestyle patterns of an advanced metropolitan nation, Australia occupies a "semiperipheral" position in the world economy.[1] An important feature of Australia's development, based historically on primary produce exports, has been the ability of successive federal governments simultaneously to appease labor (through high wages), local capitalists (via protectionist policies justified by infant industry arguments), and farmers (through a mantle of subsidization, technical support, and marketing arrangements). As Australia has become increasingly integrated into global circuits of capital, the impact of investment decisions by transnational capital and the state, in its efforts to reduce balance-of-payments deficits, is leading to a further deterioration in Australia's semiperipheral status.

We thank Phil McMichael and Raymond Jussaume for helpful comments on a draft of this chapter.

[1] There is considerable debate about what constitutes the semiperiphery (see Martin 1990). It is usually accepted that semiperipheral nations occupy an intermediate place in a global network of unequal exchange, obtaining few benefits in their economic dealings with the metropolitan core but receiving net benefits when they engage in exchange with the periphery (see Arrighi 1990). These semiperipheral nations play an important yet subordinate role in world capital accumulation. Though their living standards may for periods mirror those of the metropoles, their economies are highly vulnerable to changes within the international marketplace. William Martin has emphasized: "The semiperiphery bears the burden of the modern Janus: facing oligopolistic and political pressures from core zones and economic competitions from the periphery below" (1990:8). Some writers (see Korzeniewicz 1990) view a country such as Australia as belonging to the metropole; others (Armstrong 1978, Clegg et al. 1980) have produced evidence that Australia has remained economically dependent during the twentieth century and exhibits most of the features of a semiperipheral nation. We share the view of the latter authors. If Australia were ever at the "perimeter of the core" (Arrighi

In the period of economic vitality from World War II to the early 1970s, environmental issues were rarely addressed in political debate. The prevailing ideology was one of economic and agricultural "development"—not only as a means of contributing to the reconstruction of a war-ravaged world, but also as a means of stimulating internal economic growth. Rural producers were encouraged to produce as much as possible utilizing the most advanced technologies. The price signals from the international market and the backing of the state overruled some of the more obvious realities of price instability and environmental degradation. The excesses of the development policies and practices of the postwar years culminated in rural economic crises from the early 1970s and the acknowledgment of profound environmental damage.

The breakdown of domestic protectionist policies, the reduction in international regulations, and the transnationalization of the global economy have created certain internal tensions in Australia. Though political parties adopt an active stance in seeking domestic and international funding to stimulate industrialization, the economy is actually being deindustrialized in line with its role in the Pacific Rim economy. As a high-cost labor region, Australia is seen to be more suited to the production of raw materials. This role has placed pressure on the primary industries—agriculture and mining—to attain the necessary overseas income to help sustain the nation's metropolitan-style standard of living.

The problem for Australia, however, is that any intensification of agricultural production is likely to have a major impact on the environment. Yet the state—diminished in power as integration with international markets and firms increases—is structurally bound to support such intensification. It would appear that at the very time metropolitan nations are showing clear signs of a movement to post-Fordist industrial strategies, Australia is being required to conform to the interests of transnational food processors by maintaining Fordist strategies of production. The growth of the hitherto unfamiliar beef feedlot industry serves as an example of the type of development required by international capital and accepted by a compliant Australian state—seeking investment as a means of offsetting balance-of-payment difficulties. Though the environment is now certainly on the political agenda in Australia, it is the forests and coastal waters that have received the greatest public concern. Agriculture continues to exploit the land. The

---

1990), it is now, more than at any time in its European history, drifting further toward the periphery.

decline of Fordist agriculture in the metropoles, and its relocation to semiperipheral nations such as Australia, might be best considered a form of environmental imperialism—the exploitation of the resources of weaker nations in the world order as the core nations move toward "clean," post-Fordist agricultural development strategies.

In this chapter the Murray-Darling Basin, a large and important agricultural region in Australia, provides a case study allowing examination of the environmental impact of changes as the nation is integrated into the Pacific Rim economy.

The Australian Economy in the Global Order

After colonization in 1788 Australia's economic development was linked closely to Britain's need for cheap raw materials as inputs to manufacturing industry (McMichael 1987). Britain's intention was to ensure growth in its own manufacturing industry and political and social stability in the colonies from which raw materials were obtained. The emergence in Australia in the early nineteenth century of a property-owning bourgeoisie utilizing the labor of an agrarian proletariat provided the basis for capital accumulation in the pastoral industry. The state—a representation of British social organization—was city-based and was both a seeker of, and a conduit for, British capital investment. After population inflows during and after the gold rushes of the 1850s, the consolidation of pastoral holdings, and the continued growth of the urban labor market, Australia developed a small manufacturing sector (Boreham et al. 1989). Although labor was well organized and challenged capital for improved pay and working conditions, it was nevertheless largely compliant (Connell and Irving 1980). It also identified with the nineteenth-century bourgeois vision of progress.

Development was a motivating economic and political catchword, associated with the Australian adage "if it moves shoot it; if it doesn't, chop it down" (see Smith and Finlayson 1988). The clearing of native vegetation and the destruction of wildlife (categorized as "vermin"), together with the widespread adoption of farming practices more suited to British than to Australian conditions, had a major impact in the first one hundred years of white occupation. By 1901, the federation year of the Australian States, land had been exploited to the extent that wheat yields were about half those recorded in earlier times and native pastures had been extensively damaged through overgrazing (Heathcote and Mabbutt 1988).

Australian staple exports continued to rise to meet the needs of Britain, while Europe—particularly after the devastation of World War I—became an important market for Australian agricultural products. Declines in the terms of trade for wool and wheat caused a progressive loss of jobs in agriculture and the use of new machinery as a means of achieving productivity gains. Pastoral and farming activities moved into the drier, less environmentally suitable, regions of the continent (Catley and McFarlane 1981). The state provided continued support for agriculture and initiated various scientific and organizational councils to both boost production and provide a basis for orderly marketing. It provided railways, electrical power, irrigation schemes, and subsidized credit (Shaw 1990).

Before World War II some 79 percent of Australia's exports were farm products (Gruen 1990). During and immediately after the war, however, Australia's traditional exports—wool and wheat—either could not be transported or were given a low-priority status by importing countries. As a consequence, surpluses rose in Australia. Politicians perceived that future problems would no doubt occur if the nation relied too heavily on agriculture, and the government implemented measures—including continued high protection for manufacturing industry, increased immigration, and the encouragement of foreign capital investment—to provide a basis for "balanced" economic development.

When conditions improved in the postwar years, there was a commitment from the British to purchase Australian agricultural products. Beef was to have a special priority, and the Australian government recognized the opportunity of developing a northern beef industry (Catley and McFarlane 1981). Such a policy was seen to have the added advantage of leaving Australia's "vulnerable" northern borders with Asia less exposed to potential aggressors (Davidson 1966). "Northern development" received the unusually high status of a government portfolio. Agricultural development, particularly the substitution of labor by capital, continued alongside the growth of urban manufacturing industries.

By the late 1960s Australia had moved away from Britain as a trading partner: the United States had become Australia's largest supplier of imported goods, and Japan its largest market for exports (Gruen 1990). Australia's manufacturing industry, however, was construed to have failed to develop successfully as an export sector. It was mineral development spurred by Japanese and U.S. capital that kept the Australian economy reasonably buoyant—event during the world oil shocks of the 1970s.

The farmers have fared the worst since the beginning of the postwar economic slump in the 1970s. Since that time the contribution of agriculture to GDP, to employment, and to export earnings has been deteriorating steadily (Lawrence 1987). Efficiency and productivity goals have been imposed on agriculture, and rural reconstruction measures have supported farm amalgamation and the removal of smaller, less-viable, producers. Farms have increased in size and decreased in numbers, with those remaining farmers harnessing the latest technologies in an attempt to achieve competitive advantage (Williams 1990). Despite these efforts, the U.S. and EC subsidization and price and market support have led to Australian producers' experiencing quite severe declines in terms of trade. Increased exposure to markets—brought about through the policies of governments from the early 1970s—has, in concert with deregulation and the removal of subsidies, placed Australian producers in a precarious position.

Because the great bulk of its output enters the world market as processed foods and fiber, Australia faces the vagaries of price fluctuation associated with climatic variability, buyer resistance, changing industry and consumer demands, and tactics by competing countries. For example, because of depressed prices for wool and wheat the typical wheat-sheep farm received an income of US$1,680 for fiscal year 1991/92. The Australian Bureau of Agricultural and Resource Economics reported a further 20 percent decline in this figure for 1992/93 (ABARE 1993). In 1991/92 the agricultural terms of trade—a measure of cost/price pressures in farming—fell by 15 percent from the previous year to one of the lowest recorded levels. With economic conditions in farming considered to be as severe as they were in the 1930s and 1890s, and with farmers having experienced a major drought due to the El Niño effect (*Australian Farm Journal,* September 1993, 34–35). It is predicted that at least one-third of Australian farmers will become economically unviable (*Bulletin* July 16, 1991). Because they are unable to obtain bank loans for expansion and face declining markets and rising costs, it is expected that the majority of this group of producers will be forced from agriculture by the year 2000. In 1993 the director of the ABARE predicted that there is "unlikely to be any letup in the severe adjustment pressures facing Australian farmers over the rest of the decade" (Fisher 1993:56).

Because markets tend to produce minimum-cost solutions (Gruen 1990), rather than optimal environmental maintenance solutions, the response of farmers during the continuing crisis of the 1980s and 1990s has been to produce ever-higher volumes for export—thereby collectively exacerbating the very market and resource use problems producers seek, individually, to overcome (Lawrence 1987). The main spillover effects of short-term decision making imposed on farmers by the realities of the global marketplace have been declining rural incomes and severe ecological destruction (Lawrence 1987, Cameron and Elix 1991, Lawrence and Vanclay 1992).

The Australian Response

There was a series of major government inquiries during the 1970s and 1980s, including the Crawford Report on structural adjustment in the Australian manufacturing industry; the Myer Report into technological change in Australia; the Campbell Inquiry into the financial system; the Jackson Report on the development of manufacturing; and, more recently, the Garnaut Report on Asian-Australian trade relations. These can be read together to provide a coherent theme: while heavily protected local manufacturing industries have fulfilled certain social goals, many industries—particularly in areas where Australia must compete with countries of low-wage labor—are uncompetitive internationally. The key to the future for Australia is seen to be the development of industries that are based on the latest technologies and that provide opportunities for value adding. Economic growth will be achieved by increasing exports from Australia's resource-rich primary sector and through development of a skill-based sector. In turn Australia will import cheap manufactures from Asia and capital from the metropolitan "core" nations (see Clegg et al. 1980).

The Garnaut Report develops this theme more fully. Ross Garnaut insists that Australia must attach itself to the "ascending" economies of East Asia and develop new initiatives, including trade liberalization, greater Asian immigration, and increased foreign investment (Garnaut 1989). Garnaut argues that during the 1965–86 period Australia's share of world exports fell from 2.1 percent to 1.4 percent, while at the same time, East Asia's contribution grew from 11 to 26 percent. This trend reflected in per capita income levels, which increased by as much as 40 (Taiwan) and 50 (Japan) times in roughly the same period (*Sydney Morning Herald,* October 23, 1989).

According to Garnaut and others who perceive that Australia's economic future lies with Asia, Australia is required to produce goods for Asian markets. Since wage levels are much higher in Australia than overseas, the necessary liberalization of the economy may result in industrial restructuring, with East Asian investors providing capital and direction. It is further argued that only by becoming more internationally competitive (via restructuring) will Australia advance its role in Asia.

The main recommendations of the Garnaut Report and of the previous trade-related inquiries have been accepted by Australian governments. They have progressively removed barriers to manufactured imports and capital on the basis that doing so will force Australian secondary industry to compete internationally.[2] More-flexible labor relations (a move from unionized collective bargaining to individual employer-employee contracts) are also viewed as essential to future economic prosperity (Thomas 1991).

The problem for Australia is that the emerging trend is one of Australia's progressively being deindustrialized (as economic activity in manufacturing declines), delabourized (in association with the introduction of labor-displacing technologies), and rationalized (through the centralization—merging—of capital) (Stilwell 1986). More than 80,000 jobs per year have been lost from the manufacturing sector since 1974. New imported technologies have reduced the workplace power of organized labor. Rationalization has led to business mergers that favor international integration of key sectors of Australian industry (Rees, Rodley, and Stilwell 1993).

In essence the transnational capital "Pacific Rim strategy" (Catley and McFarlane 1981) has specific requirements: the Pacific-based center nations, Japan and the United States, are to provide capital and technology; Australia, Canada, and New Zealand are to deindustrialize while concentrating on the delivery of cheap foodstuffs, fiber, and energy; the newly emerging industrial powers, Taiwan, Singapore, and South Korea, are to produce manufactures (including items such as cars, clothing, and footwear) with cheap labor (see Rees, Rodley, and Stilwell 1993). What is of particular importance here is the extent to which the clearly defined role of Australia as a provider of raw materials for East Asia runs counter to the development of environmentally sound domestic agricultural practices.

The Current Agricultural Environment in the Murray-Darling Basin

Australia has a vast array of production regions and commodity types, and therefore it is somewhat inappropriate to discuss environmental problems in the context of a homogeneous Australia. One im-

[2] The structure of trade provides the basis for problems in Australia's current account. In 1986/87 Australia recorded trade surpluses in food, wool, iron ore, and fuels and deficits on

portant "natural" region is the Murray-Darling Basin (MDB). The Murray-Darling Basin is a natural drainage basin comprising much of inland southeastern Australia. It lies west of the Great Dividing Range and includes significant parts of the states of New South Wales and Victoria and smaller sections of Queensland and South Australia. It comprises over 11,000 kilometers of waterways, making it in aggregate terms the fourth largest river system in the world. Draining an area of over 1 million square kilometers, the basin is known for its flatness and for its great variability in water flow. The basin is Australia's most important agricultural region; it is also a region that is experiencing severe environmental degradation.

Basin agriculture, producing one-third of the nation's output, is based predominantly on wheat cropping, the open grazing of sheep and cattle, and intensive horticulture (confined to irrigation areas). The majority of farms are family-owned-and-operated, with annual production of basin farms on the order of US$8 billion (Johnston 1993:22).

This achievement has taken its toll. The soils of the basin are nutrient-deficient, thin, and easily damaged by floods, droughts, and agricultural practices such as continuous cultivation (Ockwell 1990). Increasing acidity and rising salty water tables are also severe problems. Yet the economic development of the basin has required the clearing of native vegetation, the damming of rivers, the heavy irrigation of vast tracts of the semiarid inland (areas that, for millennia, received an average annual rainfall of less than 10 inches), and the introduction of destructive foreign plants and animals.

Soil formation rates in the basin are so low that the notion of an "acceptable loss" is rejected by scientists (see Smith and Finlayson 1988). Estimates of soil loss in 1988 were that for each metric ton of grain produced, some 13 metric tons were either blown or washed away (O'Reilly 1988; and see Johnston 1993). Between 40 and 60 percent of farmers in the basin are considered to employ inadequate on-farm measures to combat soil erosion on their properties (see Lawrence and Vanclay 1992).

Salinity is another major problem (see Johnston 1993). Both irrigated and nonirrigated (dryland) areas of the basin have experienced severe salting. Salinity in irrigation areas is associated with rising water tables

---

manufactures, machinery, and transport equipment. As Evan Jones noted: "This structure of trade is characteristic of what is often called 'semi-peripheral economies.' There is a heavy dependence on primary and resource-based exports, with higher value added commodities concentrated in the pattern of imports. [For the 1980s] a wide range of 'value added' imports had increased faster than the growth in GDP and in exports of food and materials" (1989:49–50). Furthermore, Australia's dependence on manufactures and machinery from abroad has caused it to run high deficits with countries such as Japan, the United States, and Britain.

due to excessive irrigation and inadequate drainage. Dryland salinity is also due to rising water tables, with modern pastures and crops using less water than the complex ecosystems they replaced. Dryland salting can also be caused by the eventual exposure of salty lower soil horizons, arising from the erosion of topsoil by overgrazing and tree clearing (Vanclay and Cary 1989). The outcome of both forms of salination is a loss of productivity that has been valued at US$80 million per year (Cook 1988). It is predicted that at these rates of salting many of the basin's most productive irrigation areas will be unable to grow fruit trees within fifty years (O'Reilly 1988).

The total cost of land degradation is calculated to be hundreds of millions of (U.S.) dollars per year (Fray 1991, Johnston 1993). The costs of other problems, such as habitat destruction, the extinction of native plants and animals, as well as the loss of water quality because of turbidity, industrial effluent discharge, sewage, and agrichemicals, are impossible to judge. In total it was estimated that some US$1.3 billion would be required to address the environmental problems in the basin in the late 1980s (see Crabb 1988). The environmental problems of the basin have continued since that time, unabated (Lawrence and Vanclay 1992).

Tree-planting schemes and other initiatives were begun in the late 1980s, with the federal government providing funding for the 1990s "Decade of Landcare." In addition there are some 1200 landcare groups in Australia (Lockie 1992). Ironically, just as the federal government has begun to provide monies for environmental improvement, state governments have endorsed development strategies that act counter to the wider goal. For example, though the federal government has initiated and funded the Billion Trees program (the planting of one billion trees by the year 2000), over a billion trees have been removed from a relatively small region of Queensland since 1985, in accordance with that state's land development strategy (Beale and Fray 1990).

Future Options: Agricultural Sustainability or Further Exploitation?

Views differ significantly on the precise causes of the environmental problems of the basin and on the policies that might be implemented to address them. The favored approach is to view past problems as having occurred because of farmers' inappropriate attitudes toward the environment, the lack of knowledge about the damage caused by agricultural practices, and the policies of past governments, which subsidized and fostered large-scale land clearing, ill-conceived irrigation projects, and excessive applications of chemicals (see Cameron and Elix 1990, Dumsday, Edwards, and Chisholm 1990, MDB Ministerial Commission 1990, Vanclay 1992). It is also generally believed that "adaptive management" schemes, including remedial action initiated by farmers and governments, will be the key to environmental sustainability in the basin (Mackay and Eastburn 1990).

What is often forgotten are the economic conditions under which market-oriented agricultural production takes place. The structural characteristics of basin farming—high production risks, a relatively fixed supply of available land, discontinuous applications of labor, low prices, and low income elasticities of demand for agricultural products (Williams 1990, and see Buttel, Larson, and Gillespie 1990)—are long-term realities and will cause continued low returns to producers. Furthermore, the global marketplace creates the structural tendency for farmers to overproduce, to have short-term planning horizons, and to disregard the long-term returns from soil conservation (Redclift 1987, Rickson et al. 1987, Buttel, Larson, and Gillespie 1990, Vanclay 1992).

In the period of low returns dating from 1986 to 1993, basin producers knowingly have made agronomically incorrect decisions as a means of ensuring short-term economic survival (*Bulletin*, July 31, 1990, 38–44). Increasing production (overgrazing and the farming of marginal lands) and minimizing the application of inputs (such as lime to prevent acidification) have direct crop-yield and environmental implications. In addition, there is a shortage of capital attributable to reduced farm income (due to low commodity prices), the lack of equity, and the reluctance on the part of farmers to borrow (and of banks to lend) in times of falling prices. These factors prevent, or inhibit, the adoption of techniques considered necessary to improve the environment. Many farmers, compelled to work off the farm for additional income to service debts, are neglecting their farms: they do not have the time to engage in the labor-intensive, land-conserving activities that are required. In the classic sense of "agricultural involution" (Geertz 1963), farmers lack the ability to change their situation or to risk experimenting with new production strategies such as conservation-farming techniques. Large-scale environmental degradation is a serious outcome of the current farm financial crisis. In a 1991 survey of 4,000 Australian farmers, 30 percent indicated that no money would be spent during the year to address environmental degradation on their farm, and a further 50 percent indicated that the poor economic outlook would limit what could be done (see Fray 1991). The MDB Community Advisory Committee conceded that "on-farm land management practices are expensive and their implementation, in this time of economic price problems, threatens the survival of the farms" (1991:23).

When price levels improve, Australian farmers—like their competitors abroad—purchase inputs designed to increase production or lower unit costs or both, rather than invest in conservation technology. New mechanical, chemical, and biological inputs drive the farm further toward specialization, including the pursuit of Fordist-style monocultural practices, which contribute to environmental degradation (see Buttel, Larson, and Gillespie 1990). Similarly, some consider that solutions such as zero or minimum tillage create their own environmental problems resulting from the increased use of herbicides (Cameron and Elix 1991, Barr and Cary 1992). What is more disturbing is that most soil conservation practices, especially the planting of native trees, are not profitable for the individual farmer, especially in times of high interest rates and when future-discounting techniques are applied (Rickson et al. 1987, Vanclay and Cary 1989, Cameron and Elix 1991, Vanclay 1992). Australian farmers do recognize soil degradation as a general problem, but the vast majority of producers reject the notion that their own soils are being degraded (Rickson et al. 1987, Vanclay 1992).

The trends toward fewer and larger farms, greater reliance on technology, and increasing specialization in production—all in the context of world overproduction—provide the foundations for the continuation of an environmentally exploitive agriculture, one that constrains efforts to attain sustainability. Michael Redclift (1987) goes further in condemning the present course of modern agriculture. He argues that sustainable systems are characterized by diversity and stability in achieving a high level of biomass. In contrast, in modern agricultural systems large quantities of biomass are achieved by the applications of high levels of artificial inputs in less-mature and less-diverse ecosystems. Redclift sees sustainability and commercial agriculture as logically incompatible.

Rather than seeking reduced-input systems to provide a sustainable basis for farming in the basin, capital and the state have become more interested in introducing new (and potentially more damaging) forms of agriculture.

New Technologies and Strategies

Because Australia is conforming to a pattern of development that will serve the interests of the growing Pacific Rim market, it will be required to increase both volume of output and the "value-added" component of that output if it desires increased income from agriculture (Department of Trade 1987, Bureau of Rural Resources 1991). In 1990/91 Australia's annual exports included US$12 billion of agricultural produce. This was converted abroad into US$64 billion of processed goods (Bureau of Rural Resources 1991:2). It is significant that half of Australia's major exporting firms are Japanese-owned and they send about US$6 billion of unprocessed food and fiber abroad in any one year.

The federal government has initiated measures to stimulate technical efficiency in agriculture and to make producers more responsive to international price signals. It not only has reduced levels of protection so as to expose the farming sector to the world marketplace, but it also has removed regulations that have been construed as retarding Australian agriculture's links with transnational agribusiness (see Lawrence 1987, Campbell, Lawrence, and Share 1991). Furthermore, as stated earlier, by eliminating many of the restrictions on capital inflow and deregulating the banking sector, the federal government has provided the opportunity for foreign capital to influence significantly the future course of agricultural development in the basin. Two important changes have been the federal government's attempts to stimulate the growth and application of agricultural biotechnologies and the development of the beef feedlot industry.

The federal government is fostering the development of "enabling technologies," which, because of the financial risks involved, might not be considered as potential economic investments by the private sector. Of forty-six new technologies that were considered to be potentially capable of transforming agriculture, genetic engineering was ranked the highest (Bureau of Rural Resources 1991). Consequently, biotechnologies are to be selectively fostered by government policies. New biotechnologies are viewed optimistically as the solution to the dilemma of stagnating productivity in agriculture. Wheat, cotton, and a variety of other crops will be "dependent upon . . . genetic manipulation technologies" and become the "basis for a sustainable agricultural system" (Bureau of Rural Resources 1991:3,17; and see Begg and Peacock 1990). Genetic fingerprinting techniques will aid the application of plant-variety-rights legislation, so that the sale of new varieties of plants can be organized and the company holding a patent on genes can "reap an appropriate return" (Bureau of Rural Resources 1991:7).

Biotechnologies are also considered to provide the greatest opportunities for vertical integration within farming industries (DPIE 1989). Vertical integration is regarded as desirable by the state because it will allow transnational capital to exert influence over Australian agriculture and thereby effect the integration of Australian agriculture into the world economy. New directions in farming will not be led by statutory marketing authorities or by other quasigovernmental bodies (as in the past), but by transnational agribusiness. According to a Department of Primary Industries and Energy document, "Anything that blurs market signals, prevents industry adaptation to price signals or has governments take unnecessary responsibilities is an impediment. . . . The increasing industrialisation of Australian agribusiness is essential, desirable and inevitable" (DPIE 1989: abstract). Accordingly, the state considers it is the agribusiness firms that understand the international marketplace and have the managerial know-how to capture new custom (DPIE 1989).

New agricultural products will develop through market-driven arrangements. Cotton is a crop that will, through advanced biotechnologies, assist Australian farmers to improve market share. A new crop for Australia in the 1950s, cotton has become the fifth major export commodity and is the fastest-growing of all agricultural industries. About 80 percent is sold each year in raw form to the East Asia region. Farms are generally large agribusiness concerns and rely heavily on extensive agrichemical inputs. The industry is dominated by corporate interests. It has no statutory marketing boards and receives no subsidies or protection (Wormwell 1990).

Beef feedlot enterprises provide another example of change with the basin. Feedlot/abattoir complexes are appearing along the river systems of the basin to take advantage of the reliable supply of water, grain, and (unfattened) cattle (cattle that can be fattened in the feedlot). Investment from Japan, Korea, Taiwan, and Singapore has been used to develop vertically integrated complexes with direct links to Asian markets. Developments initiated by firms such as Mitsubishi, Marubini, Nippon Meats, and Itoham included feedlots of up to 60,000 head. The Riverina region of the basin, the area immediately northeast of the Murray River, is an area traditionally known for its broadacre cropping and extensive grazing; since 1988, however, feedlots with a total annual production capacity of 250,000 head have commenced operation.

With liberalization of the Japanese beef market, lot-fed beef exports are expected to continue to rise throughout the 1990s (Senate Standing Committee 1992). Basin farmers will be issued with contracts to supply grain and unfattened animals to the new complexes. Like feedlotting, contract agriculture is new to the basin (*Australian Farm Journal*, May 1991, 85), and farmers who have lost the protection of marketing boards or other support are expected eagerly to seek involvement with the feedlots. According to the executive director of the Lotfeeders' Association, "Feedlots are going to change the face of the Riverina region, creating a new economy based on supplying grain and cattle to the feedlot industry" (*Land,* January 17, 1991, 10).

Labor relations in the agricultural sector are also being targeted for change. Workers in the feedlot/abattoir complexes are expected to accept "more flexible and internationally competitive labor arrangements and awards" (DPIE 1989:67). In 1991 the first nonunion contract working team was employed in an Australian meatworks with the support of the farmers and the National Farmers' Federation. This event was in contrast to the high levels of unionization normally experienced in Australian workplaces. Furthermore, in view of the financial distress of Australian agriculture, rural workers are being required to place rural community interests ahead of union loyalty (*Australian Farm Journal,* May 1991, 38–39). There is large-scale retrenchment from rural-based industry, with nonlocals migrating to urban areas. The remaining workers tend to be farmers working off the farm or those with a farming background. They often do not share traditional blue-collar union ideology, avoid union membership, and are therefore vulnerable to structural adjustment in the industrial workplace.

Environmental Impact

Although rural restructuring raises important questions about the effects of change on rural communities and about the overall structure of the newly emerging agricultural economy, what appears to have been ignored by those supporting change is the impact of restructuring on the environment. The development and release of the products of genetic manipulation may have profound and deleterious effects on the basin's agriculture. There are concerns about ecological damage from new organisms, the use of greater volumes of herbicides (as commercial crops are modified to be resistant to proprietary products), increased pesticide applications with consequent damage to aquatic and terrestrial ecosystems, and increasing costs to farming as agrichemical companies develop biotechnological seed and chemical packages (ACA 1990, Hindmarsh 1992, Lawrence 1993, Nature Conservation Council of NSW 1993).

Developments in the cotton industry (agronomic and biotechnological) are likely to intensify existing environmental problems. For example, although cotton may be one of the "glamour" crops (Wormwell 1990), it is heavily dependent on agrichemicals and is known to have caused quite serious downstream pollution (Lawrence 1987).

Of perhaps greater significance is the development of feedlots. One estimate is that effluent from a feedlot of 40,000 head of cattle is equivalent to that produced by a city of 500,000 people. Cities of this size require the building of waste-treatment works on the order of US$80 million. In contrast, the traditional method of treatment of feedlot effluent in the basin is to contain liquid in holding ponds and to dry manure in the sun and then sell it to district farmers (*Land,* January 17, 1991). Expansion of feedlotting is likely to result in seepage and runoff that will eventually reach the already polluted Murrumbidgee River (see *Narrandera Argus,* August 21, 1990), considered the "lifeblood" of the Riverina. A State Pollution Control Commission's negative assessment of beef feedlot complexes along the inland waterways was ignored by the state government, which gave approval for their development (*Murrumbidgee Irrigator,* February 22, 1991). Even as problems of overgrazing and overcropping associated with conventional agriculture have caused havoc in the basin, the replacement of pastureland with cropland to supply grain for the feedlots may intensify current environmental problems. The potential for erosion is higher on cropping land than on pastureland (see Heathcote and Mabbutt 1988, Buttel, Larson, and Gillespie 1990), and a greater volume of fertilizers and agrichemicals will inevitably be used.

Fordism, Post-Fordism, and the Pacific Rim

It would appear that new options for basin agriculture do have the capacity to stimulate higher levels of production—but at the further expense of an already seriously degraded environment. The assurances of the federal government and scientists that biotechnological applications and new management regimes will produce a more sustainable agriculture (see Begg and Peacock 1990, Bureau of Rural Resources 1991) will undoubtedly prove specious (see Hindmark 1992, Lawrence 1993). The question is, why has Australia so readily adopted a role in Pacific Rim development that seems to ensure that its economic growth is contingent upon acceptance of further environmental degradation?

The answer appears to lie in the movement toward a post-Fordist economic structure in metropolitan nations and the exporting of environmentally harmful Fordist production strategies to the semiperiphery and periphery as the metropolitan nations adopt stricter environmental regulation for themselves. In an attempt to maintain their position in the world economy, semiperipheral nations are driven by the need to balance trade deficits. To achieve this balance, they usually rely on external capital investment and are therefore compliant in the acceptance of environmentally damaging Fordist production methods. The net effect is to worsen lifestyle conditions and to increase the economic dependency of the semiperiphery.

Fordism has been associated with a system of mass production based on the development and sale of standardized commodities to undifferentiated national markets. Motor vehicles, petroleum, and electronics were the key elements of a system that fostered productivity increases in industry, as well as provided a social-democratic system of regulation that ensured widespread consumption of mass-produced items. Full employment was a social goal of the trade union movement—a powerful agent in the Fordist regime and responsible for shaping the welfare state (see Roobeek 1987, Buttel and Gillespie 1991, Hampson 1991). Rising wage levels, which occurred in tandem with productivity increases, mitigated tendencies toward underconsumption and falling profits (Hampson 1991).

According to French regulation school proponents (see, for example, Lipietz 1987), the regime of capital accumulation, which produced the "golden age" of Fordism after World War II began to founder in the 1970s (Sauer 1990). The causes of instability during this time were significant oil price increases, the collapse of the once-stable Bretton Woods agreement, growing levels of inflation, and the transnationalization of the economies of nation-states—associated with global domination by the corporate sector (see Marsden, Lowe, and Whatmore 1990).

The technological opportunity for reorganizing production was, from the 1970s, based largely on the computerization of industry. Instead of requiring large factories with relatively unskilled workers using heavy machinery, computerized systems provided opportunities for production flexibility. For example, shorter production runs became possible and improved the capacity of firms to move quickly from one product to another. Skilled workers using sophisticated computerized equipment provided a competitive base for new industries that could identify and readily serve "niche markets" (see Piore and Sabel 1984).

Although the mass production (Fordist) and flexible specialization (post-Fordist) dichotomy has been viewed critically by theorists (see Williams et al. 1987, Foster 1988, Sayer 1989, Gartman 1991, Hampson 1991, Gahan 1993), it has nevertheless become an important distinction in the understanding of contemporary agrarian social change (Kenney et al. 1989, Friedmann 1991, Goe and Kenney 1991). From as early as the 1930s, and particularly from World War II up to the 1970s, farmers in the advanced economies—producing largely undifferentiated products and consuming industrial inputs—have been progressively integrated into the circuits of international capital (Kenney et al. 1989). Commercial family-farm agriculture became dependent on the products of mass production, such as the tractor and other agricultural machinery, as well as on fertilizers and chemicals. Price supports and a variety of other welfare-state initiatives (including taxation concessions, input subsidies, and commodity disposal mechanisms) provided farmers with conditions for stable production. The working classes of the advanced economies were advantaged by cheaper foods, which were mass produced in much the same manner as industrial household goods. New forms of management and distribution, as embodied in fast-food restaurants, enabled agricultural products to be commoditized in ways that ensured uniform quality and competitive pricing (see Kenney et al. 1991).

According to Harriet Friedmann (1991), the Fordist food regime enabled a series of commodity chains—particularly those uniting farmers to consumers—to develop and link into one of three agro-food complexes: wheat, livestock/feed, and durable foods. The livestock/feed complex is of greatest significance in the Fordist diet. In the United States extensive livestock production had been replaced by intensive methods designed to standardize meat production and to take advantage of increasingly cheaper grain feeds (Friedmann 1991). Pigs, cattle, and poultry were enclosed in increasingly smaller areas and fed grains and supplements that were standardized to produce the highest possible grain-to-meat conversion ratios. The cattle feedlot and associated abattoir became a prototypical example of "factory farming" in metropolitan countries such as the United States, Britain, and Germany. Meanwhile, nations such as Australia and Argentina—areas of cheap grazing lands—continued to produce meat for the growing global mass market in hamburgers, frankfurters, and canned-meat products (Friedmann 1991).[3] The growth of intensive livestock production in center countries was premised on the extension of the "American diet"—first to Europe, and later to Japan and the newly industrializing countries of the Pacific Rim (Kenney et al. 1989).

With the significant structural problems that the U.S. economy faced in the 1970s, the Fordist production regime in agriculture was under-

[3] Fordism takes its name from production strategies based on standardized products for a mass market, using largely unskilled workers on semiautomatic assembly lines (Gartman

mined. Martin Kenney et al. (1989) explain that the marketplace for foods has become fractured as mass-consumption diets have given way to middle-class interests in ethnic foods, chemical-free foods, and "healthy" foods. Friedmann noted also that there is a growing class differentiation in the diets of the advanced nations: "While privileged consumers eat free-range chickens prepared through handicraft methods in food shops, restaurants, or by domestic servants, mass consumers eat reconstituted chicken foods from supermarket freezers or fast-food restaurants" (1991:86). Although standardized and highly processed foods remain a key element in global food distribution, the metropolitan nations are experiencing—as part of the crisis in Fordism—rejection of the very techniques, methods, and products that so successfully tied food production to consumption in the postwar years.

Capital no longer supplies integrated national markets. It operates globally to supply regional and enclave markets using a mixture of Fordist and post-Fordist production regimes. With transnational capital having the overall say in the form and location of production, it is capable of orchestrating global production to take account of new consumer demands or profit-making opportunities.

Another essential, but inadequately discussed, part of the story is the growth of "green" movements in metropolitan nations. There is an emerging consensus in these nations that past agricultural practices are incompatible with food quality and environmental safety. For example, Hirsh and Roth have argued that "the dynamics of the Fordist reproduction process leave in their wake a progressive scale of ecological destruction" (quoted in Sauer 1990:269). This result, in addition to animal-rights violations, is one of the main reasons for European resentment of factory-farming methods. Environmentalists in a country such as Germany have introduced both ethical and cultural arguments in

---

1991). Thus, feedlotting changes the nature of agricultural employment, as it requires a largely unskilled workforce and reduces the autonomy and flexibility of farmers who provide grain and source stock. Consequently, in terms of production strategies, feedlotting would seem to be Fordist, rather than post-Fordist.

There is some confusion over what constitutes Fordist agriculture. For example, in regard to feedlot beef, Harriet Friedmann (1991) considers that intensive animal production, combined with feedgrains, forms part of a Fordist food regime, in that it produces an industrial beef product based on mass-production techniques and destined for a mass market. But Friedmann also argues that extensive beef production, which provides meat for hamburgers and hot dogs (a mass market), can be distinguished from intensive beef production, which provides meat to "privileged consumers" (niche markets). Feedlot beef in Australia is produced largely for the Asian restaurant trade. With the recent rapid expansion of the beef feedlot industry, production of marbled beef has increased enormously, the price has dropped dramatically, and, consequently, it is becoming increasingly less elitist. Therefore, though it may have been arguable that marbled, or "Kobe," beef was post-Fordist a few years ago, this is no longer the case.

questioning high-tech agriculture and its future. Thus animal husbandry is no longer something to be left to farmers—it must be guided by ethical principles and allow the interests of consumers and environmentalists to be placed alongside those of the producer (Sauer 1990). In rejecting factory farming, German consumers are providing direct and explicit support for family-farm-based reproduction strategies and environmental security (Sauer 1990).

Sweden too has initiated moves to strengthen environmental protection legislation and to provide financial support for family-farm units. This effort has been interpreted as both a challenge to mass-production agriculture and evidence of the "greening" of agricultural policy in Europe (see Chapter 2 and Vail 1991). Moves are afoot to assist producers to convert cropland to pasture, and there is an associated move to produce grass-fed beef, away from intensive forms of meat production.

Frederick Buttel (1992) has described the process of incorporating green considerations into the economic, social, and political policies of the state as "environmentalization." He anticipates that the process of environmentalization—which embodies resource conservation, sustainable development, and social justice elements—has the capacity to challenge the bases of technocratic productivist methods and ideologies.

In the United States, agriculture has been central in the debates about environmental pollution (see Kenney et al. 1989). As in the case of feedlotting, where there is a growing awareness of ecological damage and of undesirable animal husbandry practices, there is likely to be capital flight to regions of the world less prepared or able to impose rigid environmental constraints on production. In addition to Richard Goe and Martin Kenney's assertion that "large-scale production of any commodity will be a low-value business, always threatening to move to places with low land values and low labour costs" (1991:152), regional and national regimes of environmental regulation will guide decisions about industry location.

In the EC (and especially Britain), where the countryside is becoming as much a "place of consumption" as a "place of production" (see Munton, Marsden, and Whatmore 1990, Marsden et al. 1993), those whose economic and social interests are in tourism, retirement, or recreation are forcing farmers to both conserve and preserve the landscape. The new dual-income "gentry," everconscious of the advantages of rural living, are capable of mobilizing and organizing the community in an effort to protect village and community life (Cloke and Thrift 1990). Environmental pollution is very much an issue of concern for these groups.

A. Lipietz (1987) argues that environmental problems arising from Fordist production strategies have been highlighted by the groups who have lobbied for better work practices and for the realignment of business interests to issues of environmental safety. And for Goe and Kenney, "as long as a lack of environmental restrictions . . . permit an adequate rate of return for mass-produced commodities, [Fordism] will continue" (1991:152). The irony is that whereas the environmental degradation of Australian farmlands would seem to necessitate radical changes in production methods, what Australia is gaining is the uncertain futures of biotechnology and the discarded Fordist production methods of the metropolitan nations.

Capital is seeking new ways of extracting economic surplus in an increasingly competitive global economy, with the impact of change being uneven and usually socially disruptive (Redclift and Whatmore 1990, Marsden et al. 1993). New arrangements between finance and industrial capital have tended to undermine decisions made by national governments. Credit finance—upon which the transnational economy is reliant—is notoriously mobile between regions and industries (see Marsden, Lowe, and Whatmore 1990). In Australia, it has facilitated most of the mergers and rationalizations that have taken place in agribusiness and has sought new products (such as cotton, feedlot beef, and tropical fruits), new regions (such as the traditional broadacre farming areas), and new methods (vertically integrated production, direct contracts with growers, intensive animal production, and biotechnological applications) to help reorganize Australian agriculture along corporate lines.

The language of post-Fordism is in fact that of finance capital and its transnational allies. Instead of specializing in broadacre cropping and grazing under a state-organized system of statutory marketing boards, Australian farmers are being told to adopt more flexible production regimes and seek niche markets for their value-added products (National Farmers' Federation 1993). The irony here is that Australian farmers have been renowned for their diversified production regimes, at the farm level if not at the national level, and for the ability of their monopoly marketing bodies to find niche markets. Furthermore, value adding for products such as meat, wheat, and wool has been notoriously difficult because of market distance and labor costs. Because of overproduction in agriculture, effective strategies were necessary just to secure markets and often involved specialized markets and production, such as the live sheep trade, the kosher butchery, and certain types of fat-enhanced, as well as lean, meat. Many of these strategies, however, have not resulted in value adding, but have served only to secure a market for certain goods in a situation of overproduction. Consequently, some of the possible means of value adding in agricultural commodities are excluded by overproduction. It is also becoming recognized that value adding will take place closer to the retail, rather than the raw material, end of commodity chains (DPIE 1989). With the removal of tariff protection for infant industries, there is little incentive for firms to move beyond simple semiprocessing activities in Australia.

The post-Fordist discourse is not all hollow rhetoric. There is evidence that those manufacturing industries remaining in Australia are exhibiting post-Fordist characteristics, particularly new production strategies, new management strategies, new technologies, and production for niche markets (see Mathews 1989, 1992; *Australian,* November 14, 1992). Agriculture, too, cannot be construed to be uniformly Fordist. Biotechnological innovations and applications, providing opportunities for more varied production regimes and specific markets, is quintessentially post-Fordist (see Goodman, Sorj, and Wilkinson 1987). Again, in post-Fordist style the new beef feedlot/abattoir complexes are seeking to use more flexible labor arrangements and produce a specific product—marbled beef so desired by the Asian restaurant trade. But the fact that the Asian restaurant trade is itself a vast decentralized (mass) market for uniform marbled beef strips makes it difficult to sustain the notion that this situation represents niche production.

It is necessary to recognize that the movement from Fordist to post-Fordist strategies will be both regionally uneven and temporally overdetermined by state policy and by existing and potential conditions of surplus extraction. There is little doubt that livestock producers in the Murray-Darling Basin, and other regions of Australia, are becoming enmeshed in corporate production relations aimed at forcing on-farm specialization (of both crops and stock) and the integration of those producers into the Fordist, transnational livestock/feed complex. Harriet Friedmann apparently concurs: "Food is no longer simply something produced by farmers and bought by consumers, but a profitable product of capitalist enterprise, transnationally sourced, processed and marketed" (1991:71).

Conclusions

Foreign capital is poised to dictate the form Australian agriculture will take. Banking interests have usually helped to reorganize agriculture to fulfill short-term profit-making goals and have endorsed intensification of farming. Transnational agribusiness—pressed by governments, consumers, and environmental lobby groups within the metropolitan nations to initiate sounder ecological practices—has turned to Australia as a location for investment. Investment in Australia is being encouraged as federal and state regulations disappear and as the nation seeks foreign capital investment to overcome balance-of-payment deficits (Lawrence and Campbell 1991).

As Frederick Buttel (1992, following Michael Redclift 1987) has argued, "debt stress" is one of the major forces driving countries (particularly those in the periphery and semiperiphery) to introduce production-boosting technologies and practices, which result, inevitably, in environmental degradation. William Martin (1990) has stressed that historical analyses of the nations within the semiperiphery have tended to indicate their failure to move toward core status. Even if we accept that Australia has been closer to the metropole than to the periphery (Niosi 1990), it might be assumed that the combined patterns of deindustrialization and agribusiness domination of family-farm agriculture are likely to push Australia closer to the periphery.

The current pattern of capital accumulation is one that Australia, as a recipient of production regimes abandoned by the center, will be forced to accept if it hopes to become part of the Pacific Rim economy. The danger is that the relationship of Australia and the Pacific Rim will come to resemble that of Mexico and the United States. Like Australia, Mexico is well suited to range-fed beef, and as Steven Sanderson has commented, "technological and capital investments in . . . [beef production] would be better spent in ecologically sustainable, low technology range management" (1989:227). Instead, the beef industry in Mexico has been made to conform to U.S. demands for grain-fed animals, irrespective of Mexico's traditional practices.

Pollution is known to increase in line with the application of intensive forms of animal production (see Redclift and Whatmore 1990, Senate Standing Committee 1992) and, in particular, with the separation of beef-raising and grain-growing activities (see Commins 1990). By locating the least desirable agricultural activities (and toxic industries in general) in less regulated countries of the periphery (see Piore and Sabel 1984) and semiperiphery, the metropolitan nations—through the transnational corporations originating from them—are engaging in a form of environmental imperialism—specifically, formal or informal control over economic resources in a manner that is advantageous for the metropolitan power at the expense of the local economy (see O'Connor 1971).

Notwithstanding the obvious and well-focused criticisms leveled at attempts to explore "peripheral Fordism" (see Cataife 1989), this analysis has sought to provide an explanation for regional changes in a nation whose self-determination is being progressively compromised. In Australia, the Murray-Darling Basin is being restructured as a food factory and effluent-disposal system for the increasingly wealthy consumers of the Pacific Rim. It is important, but not surprising, that the Australian government is unwilling to impose tighter controls for fear of driving away much-needed capital investment. Indeed, many of the state and local governments have a vested interest in ensuring that growth is not hampered by the imposition of tighter controls.

Embedded within the changing structure of world capitalism, Australia is hostage to decisions made by those whose international economic power can not only influence—but effectively can determine—the structural character of local agriculture. There is little evidence to counter the view that farming will be considered as a convenient means of obtaining cheap inputs to the Pacific Rim's burgeoning consumer market for meats and for inputs to the food- and fiber-processing industries. The increasing level of subsumption of Australian farms with respect to agribusiness parallels the increasing level of subservience of the entire Australian economy to transnational capital (see Jones 1992). The effects on agriculture are the economic marginalization of many farmers and the continued degradation of Australia's environment.

References

Armstrong, Warwick. 1978. "New Zealand: Imperialism, Class, and Uneven Development." *Australian and New Zealand Journal of Sociology* 14(3): 297–303.

Arrighi, G. 1990. "The Developmentalist Illusion: A Reconceptualisation of the Semi-Periphery." In *Semiperipheral States in the World Economy,* ed. W. Martin. Westport, Conn.: Greenwood.

Australian Bureau of Agricultural and Resource Economics (ABARE). 1993. *Agriculture and Resources Quarterly* 5(1): 20–21.

Australian Consumers Association (ACA). 1990. *Submission to the Inquiry into Genetically Modified Organisms.* Sydney: ACA.

Barr, Neil, and John Cary. 1992. "The Dilemma of Conservation Farming: To Use or Not Use Chemicals." In *Agriculture, Environment, and Society,* ed. G. Lawrence, F. Vanclay, and B. Furze. Melbourne: Macmillan.

Beale, B., and P. Fray. 1990. *The Vanishing Continent.* Sydney: Hodder and Stoughton.

Begg, Jim, and Jim Peacock. 1990. "Modern Genetic and Management Technologies in Australian Agriculture." In *Agriculture in the Australian Economy.* 3d ed., ed. D. Williams. Melbourne: Sydney University Press.

Boreham, Paul, Stuart Clegg, Mike Emmison, Gary Marks, and John Western. 1989. "Semi-Peripheries or Particular Pathways: The Case of Australia, New Zealand, and Canada as Class Formations." *International Sociology* 4(1): 67–90.

Bureau of Rural Resources. 1991. *Strategic Technologies for Maximising the Competitiveness of Australia's Agriculture-Based Exports.* Paper no. 1P/2/91. Canberra: Bureau of Rural Resources.

Buttel, Frederick. 1992. "Environmentalization: Origins, Processes, and Implications for Rural Social Change." *Rural Sociology* 57(1): 1–27.

Buttel, Frederick, and Gil Gillespie. 1991. "Rural Policy in Perspective: The Rise, Fall, and Uncertain Future of the American Welfare-State." In *The Future of Rural America,* ed. K. Pigg. Boulder, Colo.: Westview.

Buttel, Frederick, Olaf Larson, and Gil Gillespie. 1990. *The Sociology of Agriculture.* Westport, Conn.: Greenwood.

Cameron, John, and Jane Elix. 1991. *Recovering Ground: A Case Study Approach to Ecologically Sustainable Rural Land Use Management.* Melbourne: Australian Conservation Foundation.

Campbell, Hugh, Geoffrey Lawrence, and Perry Share. 1991. "Rural Restructuring in Australia and New Zealand." Paper presented at Socialist Scholars Conference, Melbourne, July 19.

Cataife, D. 1989. "Fordism and the French Regulation School." *Monthly Review* 41(1): 40–44.

Catley, Robert, and Bruce McFarlane. 1981. *Australian Capitalism in Boom and Depression.* Sydney: Alternative Publishing Cooperative Ltd.

Clegg, Stuart, et al. 1980. "Re-structuring the Semi-peripheral Labour Process: Corporatist Australia in the World Economy?" In *Work and Inequality: Workers, Economic Crisis, and the State,* ed. Paul Boreham and Geoff Dow. Melbourne: Macmillan.

Cloke, Paul, and Nigel Thrift. 1990. "Class and Change in Rural Britain." In *Rural Restructuring: Global Processes and Their Responses,* ed. Terry Marsden, Philip Lowe, and Sarah Whatmore. London: Fulton.

Commins, Patrick. 1990. "Restructuring Agriculture in the Advanced Societies: Transformation, Crisis, and Response." In *Rural Restructuring: Global Processes and Their Responses,* ed. Terry Marsden, Philip Lowe, and Sarah Whatmore. London: Fulton.

Connell, Robert, and Terry Irving. 1980. *Class Structure in Australia.* Melbourne: Longman.

Cook, Peter. 1988. "Stewardship of Our Natural Resources: A Shared Responsibility." In *Proceedings of the First Community Conference of the Murray-Darling Basin Ministerial Council's Community Advisory Committee,* Department of Primary Industries and Energy. Canberra: Australian Government Publishing Service.

Crabb, Peter. 1988. "Managing Water and Land Use Inter-State River Basins." Discussion paper. Macquarie University, Sydney, Australia.

Davidson, Bruce. 1966. *The Northern Myth.* 2d ed. Sydney: Angus and Robertson.

Department of Primary Industries and Energy (DPIE). 1989. *International Agribusiness Trends and Their Implications for Australia.* Canberra: Australian Government Publishing Service.

Department of Trade. 1987. *Agribusiness: Structural Developments in Agriculture and the Implications for Australian Trade.* Canberra: Department of Trade.

Dumsday, Rob, Geoff Edwards, and Anthony Chisholm. 1990. "Resource Management." In *Agriculture in the Australian Economy.* 3d ed., ed. D. Williams. Melbourne: Sydney University Press.

Fisher, B. 1993. "Prospects for Australian Commodities." *Agriculture and Reserves Quarterly* 5(1): 54–59.

Foster, J. 1988. "The Fetish of Fordism." *Monthly Review* 39(10): 14–33.

Fray, Peter. 1991. "On Fertile Ground?" *Habitat Australia* 19(2): 4–8.

Friedmann, Harriet. 1991. "Changes in the International Division of Labor: Agri-Food Complexes and Export Agriculture." In *Towards a New Political Economy of Agriculture,* ed. William Friedland et al. Boulder, Colo.: Westview.

Gahan, Peter. 1993. "Matthews and the New Production Concepts Debate." *Journal of Australian Political Economy* 31 (June): 74–88.

Garnaut, Ross. 1989. *Australia and the North East Asian Ascendancy: Report to the Prime Minister and the Minister for Foreign Affairs and Trade.* Canberra: Australian Government Publishing Service.

Gartman, D. 1991. "The Aesthetics of Fordism." Paper presented at the Annual Meeting of the American Sociological Association, Cincinnati, Ohio, August 23–27.

Geertz, Clifford. 1963. *Agricultural Involution.* Berkeley: University of California Press.

Goe, Richard, and Martin Kenney. 1991. "The Restructuring of the Global Economy and the Future of U.S. Agriculture." In *The Future of Rural America,* ed. K. Pigg. Boulder, Colo.: Westview.

Goodman, David, Bernardo Sorj, and John Wilkinson. 1987. *From Farming to Biotechnology: A Theory of Agro-Industrial Development.* Oxford: Basil Blackwell.

Gruen, Fred. 1990. "Economic Development and Agriculture since 1945." In *Agriculture in the Australian Economy.* 3d ed., ed. D. Williams. Melbourne: Sydney University Press.

Hampson, Ian. 1991. "Post-Fordism, the French Regulation School, and the Work of John Matthews." *Journal of Australian Political Economy* 28 (September): 92–130.

Heathcote, R., and J. Mabbutt, eds. 1988. *Land, Water, and People.* Sydney: Allen and Unwin.

Hindmarsh, Richard. 1992. "Agricultural Biotechnologies: Ecosocial Concerns for a Sustainable Agriculture." In *Agriculture, Environment, and Society,* ed. Geoffrey Lawrence, Frank Vanclay, and Brian Furze. Melbourne: Macmillan.

Johnston, Trevor. 1993. "The Murray-Darling Basin: Inland Lifeline or Ecological Disaster." *Australian Rural Times* 3(3): 22–25.

Jones, Evan. 1989. "Australia's Balance of Payments: Recent Trends." *Journal of Australian Political Economy* 24 (March): 134–43.

——. 1992. "Multi-national Companies and the Balance on Current Account." *Journal of Australian Political Economy* 30 (December): 61–90.

Kenney, Martin, Linda Labao, James Curry, and Richard Goe. 1989. "Midwestern Agriculture in U.S. Fordism: From the New Deal to Economic Restructuring." *Sociologia Ruralis* 29(2): 131–48.
——. 1991. "Agriculture in U.S. Fordism: The Integration of the Productive Consumer." In *Towards a New Political Economy of Agriculture,* ed. William Friedland et al. Boulder, Colo.: Westview.
Korzeniewicz, R. 1990. "The Limits to Semi-Peripheral Development: Argentina in the Twentieth Century." In *Semiperipheral States in the World Economy,* ed. W. Martin. Westport, Conn.: Greenwood.
Lawrence, Geoffrey. 1987. *Capitalism and the Countryside: The Rural Crisis in Australia.* Sydney: Pluto.
——. 1993. "Agriculture and Biotechnology: Prospects for Sustainability." Paper presented to Ecopolitics VII Conference, Griffith University, July 2–4.
Lawrence, Geoffrey, and Hugh Campbell. 1991. "The Crisis of Agriculture." *Arena* 94 (Autumn): 103–115.
Lawrence, Geoffrey, and Frank Vanclay. 1992. "Agricultural Production and Environmental Degradation in the Murray-Darling Basin." In *Agriculture, Environment, and Society,* ed. Geoffrey Lawrence, Frank Vanclay, and Bran Furze. Melbourne: Macmillan.
Lipietz, A. 1987. *Miracles and Mirages: The Crisis in Global Fordism.* London: Pluto.
Lockie, S. 1992. "Landcare: Before the Flood." *Rural Society* 2(2): 7–9.
Mackay, N., and D. Eastburn, eds. 1990. *The Murray.* Canberra: Murray Darling Basin Commission.
McMichael, Philip. 1987. "State Formation and the Construction of the World Market." *Political Power and Social Theory* 6: 187–237.
Marsden, Terry, Philip Lowe, and Sarah Whatmore. 1990. "Introduction: Questions of Rurality." In *Rural Restructuring: Global Processes and Their Responses,* ed. Terry Marsden, Philip Lowe, and Sarah Whatmore. London: Fulton.
Marsden, Terry, Jonathan Murdoch, Philip Lowe, Richard Munton, and Andrew Flynn. 1993. *Constructing the Countryside.* London: UCL Press.
Martin, William. 1990. "Introduction: The Challenge of the Semiperiphery." In *Semiperipheral States in the World Economy,* ed. William Martin. Westport, Conn.: Greenwood.
Mathews, John. 1989. *Tools of Change.* Melbourne: Pluto.
——. 1992. "New Production Systems: A Response to Critics and a Reevaluation." *Journal of Australian Political Economy* 30 (December): 91–128.
Munton, Richard, Terry Marsden, and Sarah Whatmore. 1990. "Technological Change in a Period of Agricultural Adjustment." In *Technological Change and the Rural Environment,* ed. Philip Lowe, Terry Marsden, and Sarah Whatmore. London: Fulton.
Murray-Darling Basin (MDB) Community Advisory Committee. 1991. *Surviving Change: Chance or Choice?* Canberra: MDB Ministerial Council.
Murray-Darling Basin Ministerial Council. 1990. *Natural Resources Management Strategy: Towards a Sustainable Future.* Canberra: MDB Ministerial Council.
National Farmers' Federation. 1993. *New Horizons: A Strategy for Australia's Agrifood Industries.* Canberra: National Farmers' Federation.

Nature Conservation Council of NSW. 1993. *Genetic Engineering*. Sydney: Nature Conservation Council of NSW.
Niosi, J. 1990. "Periphery in the Center: Canada in the North American Economy." In *Semiperipheral States in the World Economy*, ed. William Martin. Westport, Conn.: Greenwood.
Ockwell, Anthony. 1990. "The Economic Structure of Australian Agriculture." In *Agriculture in the Australian Economy*. 3d ed., ed. D. Williams. Melbourne: Sydney University Press.
O'Connor, James. 1971. "The Meaning of Economic Imperialism." In *Readings in U.S. Imperialism*, ed. K. Fann and D. Hodges. Boston: Porter Sargent.
O'Reilly, David. 1988. "Save Our Land!" *Bulletin* 2 (August): 83–87.
Piore, M., and C. Sabel. 1984. *The Second Industrial Divide*. New York: Basic Books.
Redclift, Michael. 1987. *Sustainable Development: Exploring the Contradictions*. London: Methuen.
Redclift, Nanneke, and Sarah Whatmore. 1990. "Household, Consumption, and Livelihood: Ideologies and Issues in Rural Research." In *Rural Restructuring: Global Processes and Their Responses*, ed. Terry Marsden, Philip Lake, and Sarah Whatmore. London: Fulton.
Rees, Stuart, Gordon Rodley, and Frank Stilwell, eds. 1993. *Beyond the Market*. Sydney: Pluto Press.
Rickson, R., P. Saffigna, F. Vanclay, and G. McTainsh. 1987. "Social Bases of Farmers' Responses to Land Degradation." In *Land Degradation: Problems and Policies*, ed. A. Chisholm and R. Dumsday. Cambridge: Cambridge University Press.
Roobeek, A. 1987. "The Crisis in Fordism and the Rise of a New Technological Paradigm." *Futures* 19: 129–59.
Sanderson, Steven. 1989. "Mexican Agricultural Policy in the Shadow of the U.S. Farm Crisis." In *The International Farm Crisis*, ed. David Goodman and Michael Redclift. London: Macmillan.
Sauer, M. 1990. "Fordist Modernization of German Agriculture and the Future of Family Farms." *Sociologia Ruralis* 30(3/4): 260–79.
Sayer, A. 1989. "Post-Fordism in Question." *International Journal of Urban and Regional Research* 13(4): 666–95.
Senate Standing Committee on Rural and Regional Affairs. 1992. *Beef Cattle Feedlots in Australia*. Canberra: Commonwealth of Australia.
Shaw, Alan. 1990. "Colonial Settlement 1788–1945." In *Agriculture in the Australian Economy*. 3d ed., ed. D. Williams. Melbourne: Sydney University Press.
Smith, D., and B. Finlayson. 1988. "Water in Australia: Its Role in Environmental Degradation." In *Land, Water, and People*, ed. R. Heathcote and J. Mabbutt. Sydney: Allen and Unwin.
Stilwell, Frank. 1986. *The Accord . . . and Beyond*. Sydney: Pluto.
Thomas, Richard. 1991. "Inequality and Industrial Relations." In *Inequality in Australia: Slicing the Cake*. Social Justice Collective. Melbourne: Heinemann.
Vail, David. 1991. "Economic and Ecological Crises: Transforming Swedish Agricultural Policy." In *Towards a New Political Economy of Agriculture*, ed. W. Friedland, L. Busch, F. Buttel, and A. Rudy. Boulder, Colo.: Westview Press.
Vanclay, Frank. 1986. "Socioeconomic Correlates of Adoption of Soil Conserva-

tion Technology." Thesis, Dept. of Anthropology and Sociology, University of Queensland, St. Lucia, Australia.

——. 1992. "The Social Context of Farmers' Adoption of Environmentally Sound Farming Practices." In *Agriculture, Environment, and Society,* ed. Geoffrey Lawrence, Frank Vanclay, and Brian Furze. Melbourne: Macmillan.

Vanclay, Frank, and John Cary. 1989. "Farmers' Perceptions of Dryland Soil Salinity." Discussion Paper, School of Agriculture and Forestry, University of Melbourne, Melbourne, Australia.

Williams, D. B., ed. 1990. *Agriculture in the Australian Economy.* 3d ed. Melbourne: Sydney University Press.

Williams, K., T. Cutler, J. Williams, and C. Haslain. 1987. "The End of Mass Production?" *Economy and Society* 163.

Wormwell, G. 1990. "Cotton Grows Glamorous." *AIM* 10 (November): 1.

PART II

SECTORAL RESTRUCTURING

4

Finance Capital and Food System Restructuring: National Incorporation of Global Dynamics

Terry K. Marsden and Sarah Whatmore

By the mid-1980s researchers had increasingly recognized that agrarian social relations were deeply embedded in the growing international restructuring of economies and societies. Writing from a largely urban and regional perspective, they applied a variety of interdisciplinary approaches to the analysis of this restructuring (Lash and Urry 1987, Cooke 1989, Martin 1989, Jessop 1990, Harvey 1991). The fundamental crisis of Fordism that they identified was related to a confrontation with certain technical and human limits that engendered slowdowns and real falls in both productivity and the rate of capital accumulation and profit (Glyn 1989). For instance, in terms of (gross) capital stock for the business sectors of the United States, Japan, West Germany, and the United Kingdom since 1973, the rate of accumulation by the mid-1980s was within the range of one-half to two-thirds of its 1973 level. This rate reflects a growth in business investment running at about 3 percent per year since 1973, less than half the growth rate of the previous period.

In addition, the profound transformations ensuing in the economic structures of advanced economies were being paralleled by a restructuring of their financial systems. It seemed that the decline in the Fordist mode of regulation was initiating far more than the steady unilinear evolution of a new productive system; the corollary was a competitive international restructuring of finance relations and their national regulation. These could condition the degree of uniformity or unevenness such new productive systems would encompass and were likely to affect the degree to which new forms of production could remain sustainable.

A focus of particular concern assessed the implications of these processes for the industrial reorganization of labor (Massey 1985) and the implications for new regional, as well as international, spatial, and social divisions of labor. In particular, as macroanalysis concentrated on the decline of U.S. industrial and financial hegemony (Lipietz 1990), it has been common to focus on the regional implications of the growth of new social and economic forms that have required different sets of spatial priorities in terms of site location and degree of interfirm linkage (for instance, flexible manufacturing systems, just-in-time delivery, more-flexible working practices). Less attention, however, has been paid to how these forces both promote and require new types of dependent relations *between* banking and industrial capital at different spatial scales of analysis.

Researching these relationships, as R. Minns (1981) suggests, is far from easy, particularly given the growing level of internationalization of both spheres and their necessary interactions in national and regional social formations (see Thrift and Leyshan 1988). Nevertheless, it is increasingly recognized that the industrial restructuring ensuing from the early 1970s in most advanced economies could be achieved only by a realignment of these structural relationships at different spatial and sectoral levels. This realization was slow to revive a Marxist debate surrounding the role of banking capital in industrial development and the extent to which the merging of these two spheres represented the development of international finance capital to a hegemonic level (McMichael and Myhre 1991).[1]

The food and agricultural sectors were in no sense immune from these tendencies. New sets of relations between banking and industrial capital, as well as their service-based ancillary activities, brought about the growth of mergers, buyouts, and asset-stripping during the 1980s, not least in the food sector. A heightened reliance on banking finance through a range of credit relations reinforced the geographical mobility of capital and the uneven disassembly of previous agro-industrial structures. The increasing tendency for the concentration of production in the food sector was related to falls in the rate of profit, due to rising

[1] We use the term *finance capital* in a broad sense denoting what G. Thompson calls "the articulation of three primary forms of capital—industrial capital, commercial capital, and banking capital" (1977:23). *Banking capital* represents both the fixed and liquid assets of the banking sector specifically, as well as their institutional organization. This banking capital is a major component of finance capital.

internationalized competition and the uneven adoption of technologies. The overall inelasticity of food demand placed an even greater burden on food manufacturers as they recognized that the real gains in market power were associated with product value-added differentiation and near-market distribution and retailing. As a consequence, there was further concentration based on high levels of credit dependence. For instance, between 1983 and 1988 the multinational conglomerate Unilever sold ninety firms (worth on the order of £2.3 billion) and purchased a further hundred (costing £4.7 billion). Supported by merchant banks, the companies' overall borrowing limit exceeded £14 million, largely because of their attempts to obtain—as with their main competitors BSN and Nestlé—established markets in Europe. G. Clark (1989) also illustrated, with the case of the leveraged buyout of RJR Nabisco by Koulberg Kravis Roberts (KKR), how such tendencies reorganized the geography of industrial management and passed on severe consequences for labor organization and workers' rights (Clark 1990).

The growing incidence of takeovers, amalgamations, and disinvestments of subsidiaries by parent companies, as well as the increasingly sophisticated financial technology employed for these purposes (for example, leveraged buyouts, share repurchases, and allied transactions), represented a shift to a new set of relations between industrial capital (of which the food sector was a significant element), finance capital, and emerging regimes of national regulation adopted to facilitate and legitimize these spheres. As T. M. Rybczynski (1989b) and Minns (1987) suggest, these developments embodied new relations between banking and industrial capital, reducing the regulatory power of the nation-state. In Britain, for instance, U.S.-based companies borrowed from both British- and U.S.-owned banks; conversely, British companies there increasingly borrowed from British-based, U.S.-owned banks, which had initially been established to finance U.S. multinationals. The Imperial Chemical Industries (ICI) group, for example, uses no less than three hundred banks throughout the world: locally established banks for local currency financing and international banks for loan finance. It also used German banks in organizing some of its late 1980s Eurobond borrowings.

These tendencies have held significant consequences for the organization of the land-based agricultural sector. They have both enabled industrial capital to overcome many of the rigidities associated with exploiting farm-based activity and established revised economic and social parameters for the regulation and flow of capital and technology to selected groups of producers and agricultural regions. Such interrelationships thus hold important implications for farm- and firm-based uneven development. Moreover, the particular national characteristics of banking systems play an important role in reshaping agricultural relations and the regulation of the farm "crisis" (see Goodman and Redclift 1989).

When faced with new rounds of uneven financial and industrial restructuring, scholars tend to construct a broad evolutionary picture stressing the diversity and complexity of current and forthcoming conditions (Hirst and Zeitlin 1991). It needs to be recognized that such evolutionary analyses should be tempered by the analysis of diverse and contingent social and economic conditions, which in themselves contribute to new rounds of broader uneven development. Globalization becomes nationally and sectorally incorporated. In this chapter, we outline the key tendencies reshaping the relationships between the finance and industrial sectors from the 1980s onward. This outline sets the parameters for the ensuing, more specific, analysis of national banking and its relationships with agro-industries and the food production sector.

Underwriting our treatment in this chapter are three premises. First, we are aware of the current problems of social theory in addressing contemporary issues concerning the competitive interface of the financial and industrial spheres and their relationships to the role of national and international state action. In particular, it is necessary to assess the ways in which globalized capital relations are nationally and sectorally incorporated. Second, and as a corollary, we are also aware of the need to link these spheres, theoretically and empirically, as mutually constituted relations, to help advance to the broader restructuring debates. Third, with respect to agrarian change, our focus suggests the need to link these abstract considerations directly to a multilevel analysis of contemporary uneven development (see Marsden, Flynn, and Ward 1991). This linkage requires a recognition of both the particular distinctiveness of the food system within the broader global economy and a focus on the points of transaction with wider social, economic, and political processes, whereby the food system represents one significant subset of overall capitalist development (Marsden and Little 1990).

Key Dynamics of Banking Industrial Relations

The restructuring of the financial and industrial capital interface is represented by several key features. They are analyzed only partially in the literature, yet they are fundamental to an understanding of the restructuring of advanced economies in the 1990s.[2] First, the period since the early 1980s represents a distinct phase in the internationalization of financial markets. During the 1970s the progressive internationalization of the money markets was dominated by the recycling of the dollar surpluses of oil-exporting countries, through the Eurocurrency markets based particularly on London and toward underdeveloped countries. During the 1980s, as a result of the Third World debt crisis and the rise of surplus capital in Japan and Germany, capital was reinvested into the large multinationals and the governments of the advanced economies (Lipietz 1990). This reinvestment fostered the growth in the Eurobond markets and, later, in the Euronotes and Euroequity markets (Coakley 1989, Evans 1989).

Changes in the geography of dominant financial flows during the 1980s also brought forth modifications to the operation of financial markets and the role of banking sectors. Securities markets (bonds and equities) rose by means of the process of *securitization,* which involves the buying and selling of financial instruments and the trading of bonds and equities in corporate restructuring. As P. Williamson and D. F. Lomax (1989) pointed out, securitization put traditional banking under considerable competitive pressure, with an aggregate decline in conventional bank lending to industry and a growth in the use of securities. One example was the growth of the "swap" market, which mushroomed in the period 1982–86, representing US$400 billion in 1986. Other mechanisms include Eurocommercial paper, Euroequities, equity-linked bonds, as well as various futures and option instruments. The rise of these forms of intermediation has been enhanced by the deregulation of financial markets progressed by certain nation-states (e.g., US, UK); but they also have tended to reduce London's hegemonic role as a preeminent financial center. Both New York and Tokyo have large domestic securities markets and houses.

The development, especially in the United States and the United Kingdom, of a more securitized financial system has heralded a period of new influence over the national and international industrial systems. For instance, in addition to wide and deep primary and secondary capital and credit markets, which cover the bulk of flows of new and old savings, two principal new elements concern the market for corporate

[2] For instance, financial-industrial relations are largely ignored in debates concerning the industrial restructuring and regulation of production and consumption, despite a recognition that, in organizational terms, the position of the finance sector is just as important within a post-Fordist set of economic arrangements.

control and that for venture capital (Rybczynski 1989a). The former is concerned with the phasing out and changing nature of past savings embodied in physical assets and, linked to it, human and financial assets. Venture capital and associated markets (leveraged buyouts) are concerned with the transfer of new savings to new firms and new industries with uneven risk characteristics. These markets can accelerate the phasing out of activities (as they did significantly in the 1980s) because of changes in comparative advantage either domestically or overseas. This transfer provides more flexibility in capital investment and is linked directly to stimulating rapid reorganization of management and labor. Such mechanisms undoubtedly change the nature of both firm-based and national-based risk and provide opportunities for the costs and rewards of restructuring to be rearranged both spatially and sectorally (Clark 1990).

More market-oriented and diversified financing systems (particularly in the UK and the US) have brought about three important effects regarding industrial restructuring. First, they have increased the capacity of a national economy to take a greater degree of risk increases. Capital and credit markets cover an increasing proportion of productive assets. They also channel the bulk of new savings in such a way that they are accessible to a greater number of savers and saving institutions, with a larger number of financial instruments covering different combinations of risks and rewards. Second, the process of monitoring and evaluating the performance of credit and financial investment is now spread generally among a larger number of providers of funds and managers of savings (*Financial Times* Special Issue, November 1991), and the pricing of expected risks and reward combinations is more comprehensive and assessable over short-term horizons. Third, as D. Vittas (1989) suggests, the selection and recruitment of managerial teams—whether associated with changes in ownership and control or not—is approached in a more competitive and professional way.

The emergence of securitized financial systems is characterized by a heavy reliance of firms on external funds raised through capital and credit markets (Cumming 1987). The financial system provides for the needs of increasingly international industrial frameworks and, in particular, has aided the concentration and corporate dominance of a smaller group of internationalized businesses (Amin and Dietrich 1990). The dominance of a few internationalized businesses had been particularly significant in the food sector in the United Kingdom. It has encouraged further internationalization of existing British-based food firms, bringing rapid internal reorganization. Though the merger boom of the mid-1980s was financed by the raising of cash (approximately 15 percent), by issuing equity (64.2 percent), and by using credit (20.6 percent), the growth in debt-based activities was the most significant historically (Scouller 1989, Corrie 1987). The 1986 boom in mergers, for example, was characterized by a particular growth in food, drink, and tobacco mergers, representing 36 percent of all merger activity. This growth was not only associated with rationalization strategies; a major motive was to expand "quality-based" market growth. These activities were increasingly associated with foreign acquisitions—expenditure for foreign acquisition by British companies increasing fourfold since the late 1970s. These included major acquisitions by Unilever (acquiring Cheeseborough-Ponds), International Chemical Industries (Glidden Paints), Boots (Flint), and British Petroleum (Purina Mills). The peaks in merger activity (in the UK in 1968, 1972, and more drastically in 1986) all coincided with peaks in stock market activity and growth in profitability of firms over the short term.

During the 1980s financiers began assembling the enormous quantities of capital necessary to fund key, large acquisitions. This accumulation of capital was also encouraged by state fiscal policy. In 1984, for instance, the British government dropped the tax investment concessions concerning accelerated depreciation. Consequently, investments in new plants and equipment were more expensive than before, and thus the purchase of up-and-running companies was more attractive. A more relaxed state attitude to large-scale mergers, as well as an ideological commitment to the policy of encouraging private-sector restructuring of the economy rather than state-directed mediation, promoted more dynamic and debt-based links between the finance sector and industry. Within the food sector, this policy encouraged further reorganization, with the entry of new arrivals to the superleague of diversified food companies.

One example is the growth of Hillsdown Holdings, now the third largest European food conglomerate. It boasted a turnover of £3 billion for 1987 and held a total of 150 subsidiary companies in Europe (including Buxted Poultry, Daylay Eggs, and Ross Poultry). As an important market shareholder in red meat, bacon, poultry, and eggs, it also supplies the majority of its nonanimal feed requirements from ten mills and its own chicks from commercial hatcheries. Downstream processing activities are integrated within the group through the control of twenty-five abattoirs and several food-processing, distribution, and meat-trading companies (Ward 1990).

As with other notable corporate acquisitions of the 1980s (e.g., the

Hanson Group's acquisition of the Imperial Group for £2.6 billion and its emerging share interest in ICI), the agglomeration of these larger internationalized holding companies depends on high levels of bank borrowing and a strategy of asset-realization to finance the acquisitions and pay debt. The flexibility of the financial system is crucial to these developments, but its very logic influences the speed and intensity of internal restructuring soon after acquisition occurs. Particularly in the context of relatively high real interest rates in the United Kingdom (see Dixon 1991), "asset-mining" (i.e., developing undervalued assets of previous hidden value) is complemented by "asset stripping" in order to allow repayment of loans, enhance collateral, and increase short-term investor confidence.

Banking and Agro-Industrial Relations in the United Kingdom

Incorporating the Nation-State

These emerging relations between the financial sector and industry are highly uneven among nation-states (Coakley 1988, Rybczynski 1990, Dixon 1991). Whereas this unevenness may represent the development of new imperialist rivalries (Rowthorn 1971), based not only on the development of capital surplus but also on the conditions each nation-state places on its financial sector, the differences also provide an uneven platform for industrial restructuring and the productive social relations that go with it. It is widely recognized that, partly because of their different historical roots, countries such as Germany, Japan, and France still retain a more distinctive banking sector without the growth of securitization to the same degree (Vittas 1990). Moreover, despite both the real and rhetorical significance of European integration in 1992, important differences exist in the degree of state financial regulation among nations (see Dixon 1991, Dickson 1991). Partly as a result, though there is extensive provision of shorter and longer term bank finance of industrial sectors (both large and small), the provision of equity and bond finance for industrial companies is limited, although it is growing, especially in Japan.

In the United Kingdom, partly as a consequence of state deregulation, the arrival of a large number of foreign banks and the growth in the use of medium-term bank loans have led to increased competition among banks for the business of the larger corporations (Lomax 1989). Multiple-banking has become prevalent, resulting in the fragmentation of corporate financing among many competing banks. Industrial corporations tend to adopt a more highly structured approach to multiple banking in order to reduce the undue proliferation of relationships. They classify banks into separate tiers, ranging from a small group of principle bankers to banks that may be invited to bid for particular services on the basis of their specialist or geographical strength. Hence, both in the nationally based methods of regulation and in the variety of ways in which banks relate to industries at the local level, there exists a considerable degree of capital mobility and growth between different places. This represents a significant mechanism in uneven spatial economic development (Green 1988). Some evidence suggests that the differences in the patterns of relations between bank-based (e.g., Germany, France, Japan) and market-based (UK and USA) systems are declining at an international level because of the current trends in globalization of capital markets and the securitization of bank lending. In the United Kingdom, however, the use of industrial experts by banks, the increasing appointment of bankers as nonexecutive company directors, and the greater willingness to take equity stakes in companies with good long-term prospects underline a closer set of interlinkages between banks and industry.

A major consequence of the new rounds of uneven development between the finance sector and industry at the national level concerns modifications in the nature of what we can call "microcredit relations." In the agro-industrial and agricultural sectors, increasing competition between credit agencies and the acquisition activities of the food industries have stimulated the growth of new credit instruments, allowing more diverse arrangements concerning repayment, risk taking, and interest bearing. In the United Kingdom, this growth is directly linked with state policy, particularly attempts to deregulate the banking sector and foster more competition, at least in the short term. The "big bang" in the stock market encouraged international bank trading and further exposed national systems of finance to international forces. Moreover, as we have seen, corporate capital has been investing heavily overseas. Indeed, multinational investments (principally based in the USA, Japan, and Europe) have been growing three times faster than world trade since 1983. As a consequence, patterns of trade, financial flows, and technology transfer are increasingly converging on the pattern of uneven foreign direct investment, which increased by 28.9 percent a year between 1983 and 1989.

An overall consequence of the deregulation in the national banking sector has been the replacement of one set of oligopolistic credit market relationships by another that has significantly weakened existing nation-state interference. This development is still unfolding with attempts to "reregulate" banking at the European level (Cerny 1991). Nevertheless, the aftermath of such regulation is very much characterized by allowing the traditional divisions (e.g., between bank or building societies, or indeed specialist sectoral credit agencies such as agricultural banks) in the market to be broken down in the name of providing a more competitive service to the consumer.[3] This breakdown has sensitized banks to differences in marginal interest rates both nationally and internationally and has placed considerable emphasis on diversifying and packaging the price of credit around sets of other banking services, for example, insurance, leasing, pensions, and tax-avoidance advice. Moreover, the Banking Act of 1987 attempted to enforce a more legally binding form of supervision of banking practice in the United Kingdom, even though banking has remained largely self-regulatory. The act has also significantly failed to reduce public anxieties about the increasing power and undemocratic practices of the finance sector.

In their search to extend their services to new national markets, banks have begun to supply credit for the purposes of public consumption as well as corporate production. In the United Kingdom and across much of northern Europe, this strategy is most noticeable in the mortgage industry associated with housing. This tendency also benefits individual firms indirectly by providing more consumers with the finance to purchase products. Banks at the national level thus reduce a critical market constraint of industrial capital. They provide mechanisms that can prize open areas of public and privatized consumption and thus allow more nationally based market opportunities for transnational industrial capital. This result has been particularly evident in the United Kingdom, as the role of nonmarket goods associated with the welfare state (such as public housing and other forms of collective consumption) has been progressively dismantled.

We can see that the restructuring of the banking sector and that of industrial capital have been mutually reinforcing in advanced capitalist societies in the late twentieth century. Both are more internationalized and consumer-oriented, and the sets of links between them are highly

[3] The traditional distinction in the United Kingdom between building societies as mortgage providers and clearing banks has been progressively broken down as a result of the Building Societies Act of 1986. Because of the blurring of this distinction, there has been more intense competition in mortgage lending and a proliferation in credit and services. During the 1990s there is likely to be a growing level of concentration among building societies and banks.

complex and pronc to breakdown (e.g., the 1987 stock market crash).[4] Moreover, they influence, in a much more diversified and succinct way, local social action. Within this somewhat vulnerable context of the restructuring of the transnational financial and industrial relations, it is thus important to focus, at a secondary level of analysis, on the operation of these processes within the nation-state boundaries. It is through such analyses that we can begin to establish how globalized capital relations are necessarily locally embedded within both particular sectors (for instance, food) and particular regions. In addition, it is necessary to position such analyses within particular phases of state-agrarian relations.

We attempt to examine these processes in more depth by linking the emerging strategies of the finance sector with agro-industrial firms and the agricultural sector at the national level. This approach illustrates the nature of the regulatory role of the finance sector both in affecting agrarian production relations and in devising mechanisms for reproducing markets that industrial capital can then exploit. The British case also demonstrates how the nation-state, partly by dint of its historical bases, influences the relationships between banking capital and agro-industrial capital.

National Banking and Agro-Industrial Relations

In the agricultural and food sectors the growth of production was supported strongly by state agencies in Britain and in other parts of northern Europe from the 1940s. In direct contrast to the policies concerning the finance sector in the United Kingdom, agriculture and food became tied explicitly to direct forms of state intervention in order to stimulate production and regulate consumption (Marsden, Flynn, and Ward 1991). Through a system of price supports and subsidies to production-oriented technology, farmers were encouraged to purchase the products of the burgeoning industrial input suppliers—feeds, grains, machinery, and fertilizers and pesticides. This system produced a reliable and, until the early 1980s, expanding market for agro-industrial firms.

[4] There is considerable debate as to the causes and consequences of the international stock market crash of 1987. An overinvestment in the securities industry has been seen as one principle characteristic that resulted in both excess capacity and overtrading in securities markets. The aftermath has witnessed a shakeout leading to further concentration and centralization within the finance sector and further consolidation of Japanese hegemony in international finance markets, displacing the U.S. banks from their leading position (Coakley 1988, Lipietz 1990).

Unlike that of many of its European neighbors, the British finance sector was left largely untouched by these protectionist and agricultural productionist policies. The banking sector would engage in lending if the collateral status of farmers was seen as sufficient and the industry as a whole was seen to have stable growth.[5] In France, by contrast, the development of a sector-specific, state-supported agricultural bank (Credit Agricole), in addition to a state-encouraged cooperative infrastructure (Clauzier 1989), established concessionary interest rates to farmers. This development was seen as a direct aid to farmers' cost-price squeeze problems, protecting the rural economy, and enabling market openings for agribusiness. Indeed, in other European countries (e.g., Greece, Spain, Germany, and Belgium), the period of state-induced agricultural productionism was typified by the establishment of agricultural banks in receipt of state finance, allowing concessionary loans to their agricultural sectors. By the 1980s, however, productivist state policies in Europe began to wane (see Chapter 2) under pressure from the EC's Common Agricultural Policy budget reductions and growing environmental concerns. In conjunction with these pressures, food and finance relations destabilized, resulting in an "international farm crisis" and severe oligopolistic struggles between, and within, the banking and agro-industrial sectors. Previous nationally variable relationships between the finance sector, agribusiness, and the state began to influence the depth and diversity of the farm crisis under the new circumstances.

These tendencies are well illustrated by reference to the British agro-supply sector. Evidence suggests that Britain holds the most concentrated food sector in Europe. Even as early as 1979, fifteen of the largest twenty-one food firms in Europe were British-based. Moreover, the agrosupply industry represents the most concentrated subsector of the food system in the United Kingdom, with fertilizer production now in the hands of two multinational firms (Norsk Hydro and Kemira). Together they have the potential to hold 90 percent of the market. Within Western Europe as a whole, the number of independent fertilizer companies fell from fifty-six in 1980 to twenty-nine in 1990. Kemura spent

[5] The only direct state-supported credit facility was provided by the establishment of the Agricultural Mortgage Corporation (AMC), which provided loan arrangements largely for the purchase of land. The AMC was established under the Agricultural Credits Act of 1928 to provide a source of capital for the agricultural industry. It is registered as a public company, its shareholders consisting of the Bank of England and the five major clearing banks. Under the 1928 act, the AMC makes loans to farmers against a first mortgage on agriculture and forestry land in England and Wales and offers improvement loans. Legislation restricts lending to these purposes.

US$257 million from 1986 to 1991 in acquisitions, more than tripling its fertilizer production capacity to 7 million tons per year. Norsk Hydro's capacity is now twice this, making it the largest European producer.

In the feedstuffs sector, BOCM Silcock (a wholly owned subsidiary of Unilever), Dalgetty and Pauls, and J. Bibby (Barlow Randt) control 57 percent of the British market. Because of their reliance on the farm sector for a considerable proportion of their raw materials as well as sales, their levels of profitability are particularly sensitive to the fluctuations in farm incomes brought about as a result of changing agricultural policy. For instance, like the agricultural machinery firms, feedstuffs manufacturers were badly affected by the introduction of milk quotas, with a fall in market demand from 4 million tons in 1983 to 2.5 million in 1987. As a result, the feedstuffs sector suffered from overcapacity and a need to reduce the number of plants (usually through a process of further acquisition and takeover).

In the agricultural machinery market a similar position prevails, with the four largest companies (Ford, Case, Massey Ferguson, and John Deere) controlling 77 percent of the British market. With the purchase of agricultural inputs representing 60 percent of all farmers' costs, any fall in farm incomes directly affects these input markets. For instance, the market for new tractors fell from 35,000 per year in the early 1970s to 19,000 by the mid-1980s. This decline stimulated a reorganization of producer plants and the franchised dealer system, with more direct control over sales operations (e.g., regular meetings, product launches and contract classes, exclusive agency arrangements). Of even more significance in protecting and opening markets has been the provision of concessionary credit packages administered by leasing companies associated with the agromachinery firms and the main clearing banks.

The period from the late 1970s onward thus represents a more unstable period for the agrosupply sector, facing both industrial restructuring processes leading to further concentration, on the one hand, and a dwindling market situation brought about by the demise of the postwar productionist agricultural policies on the other.

The Role of Credit Agencies in Regulating Agricultural Retrenchment

A principal indicator of the economic entrenchment of the nation's agricultural market concerns the fall since 1980 of the real level of fixed capital stock in agriculture. This level had risen steadily since 1945, but declined in real terms by 5 percent between 1980 and 1991, suggesting that the aggregate increase in capital intensity had been checked and, in some cases, reversed. This trend has been particularly noticeable in plants and machinery, where farmers are far less willing to replace equipment rapidly. In addition, and of particular importance concerning lending institutions requiring security of assets as a condition for extra lending, land prices began to fall (by 25 percent since 1987). Although land and buildings remain the most important assets on farms, their value became more unstable during the 1980s, representing 76 percent of all assets in 1975, but only 66 percent in 1986. The increasing instability of the farm production sector further encouraged a restructuring of the supply and demand of credit and influenced qualitative changes in the nature and longevity of credit relations themselves. With the internally generated process of restructuring operating in both the financial and agro-industrial spheres, the gradual demise of the agricultural productionist order provided an additional critical conjunction. It further reinforced the domination of oligopolistic industrial and finance capital over the less concentrated farm production sector. It influenced the strategies of both the banks and the agrosupply sector toward agriculture.

Between 1976 and 1989 in the United Kingdom, total farm debts rose from £2,400 million to £9,750 million, which in debt-to-asset ratios represents an increase from 9.9 percent to 16.36 percent by 1989. When the value of fixed assets is excluded from the debt-to-asset ratio, the proportion increased from 28 percent in 1978 to 56.8 percent in 1989. If one compares debt liabilities with equity and income a similar trend is found, with an increase in debt liabilities representing an equivalent of 8.8 percent in 1972 to 19.32 percent in 1989. By 1989 the British agricultural credit market stood at £9,750 million, representing more than a twofold increase since 1980, with only four main clearing banks holding over 70 percent of all liabilities. The remaining 30 percent is absorbed by the Agricultural Mortgage Corporation, industry, and individuals. The four clearing banks are responsible for 95 percent of total bank lending to agriculture, with 30 percent of lending medium- and short-term loans and 70 percent overdrafts. The agricultural portfolios of the four are relatively small, with agricultural lending representing only 3 percent of total lendings at Lloyds and 3–5 percent at Barclays. This proportion declined from 8 percent in 1973. Thus, unlike the agro-industrial sector, any further deepening of the farm crisis in thc United Kingdom was likely to inflict only marginal damage on the British banking system.

This system differs very much from that in the United States and in other European states. For instance, in France the Credit Agricole had, until 1990, held a monopoly over agricultural lending through the issue of state-supported cheap loans to the agricultural sector. Since 1990, however, the state-supported loans have become much more restricted and, through a policy of deregulation, the Credit Agricole no longer retains its monopoly on agriculture or on the distribution of "bonus loans." As a result of these changes it is likely that the banking relationships with farmers in France and the United Kingdom are likely to converge, with less emphasis being placed on supporting the less productive and less profitable farm sectors. Four other banks, Société Générale, Credit Mutuel, Banque Populaire, and Bank Nationale Paris, can now compete to provide cheaper loan arrangements to farmers. In addition, in the context of further European integration of capital markets, the opening of a branch of the powerful Dutch Rabobank in Paris in 1989 suggests increasing internationalized competition in agricultural lending (see also Kaye 1989, Dixon 1991). These banks are now developing more closely defined loan packages to farmers for specific types of agricultural investment.

In the British case, with interest rates higher and somewhat more volatile because of the lack of state direction, the most rapidly expanding credit agencies are the leasing companies owned by the major agro-input manufacturers. These companies are tied to both industrial and banking capital and multiplied their level of lending tenfold between 1983 and 1986, accounting for £400 million in 1986. The rapid rise in farm debts in the early 1980s caused the clearing banks to review lending strategies to the agricultural sector. This review involved revising what previously had been blanket policies of lending (using principally the overdraft facility on the basis of security against assets) to more specific lending related to business' profitability and prospective viability. Historically, in Britain agricultural businesses were given no special status in the provision of credit in comparison with other businesses. The ownership of the land assets was seen as sufficient collateral for releasing overdrafts and loans. By the late 1980s, however, all banks were encouraging farmers to restructure their core overdrafts to more tightly defined loans and "more efficient cash use." The producer was required to monitor the day-to-day management of finances and forced to question whether borrowing was required over specific time periods (Marsden, Lowe, and Whatmore 1990).

In addition to these constraints, the banks, in collaboration with leasing companies, provide specific point of sale credit (i.e., a delivery system that ties borrowers to specific purchases and facilitates the sales of manufacturers' and merchants' agro-inputs) and leasing schemes—where equity is removed from the farmer and the bank directly pays the supplier, with the farmer paying off the lease to the bank. Approximately 60 percent of all new tractors are now "sold" in this way.[6] More specifically, the lending strategies of the four main clearing banks in the United Kingdom can be divided into two broad categories. First, both Barclays and National Westminster have developed specialist credit packages and delivery systems specific to agriculture. Second, Lloyds and Midland tend to gear their general credit facilities to the farm sector without any specific agricultural focus. All four banks are placing increasing emphasis on the production of budgets, balance sheets, and cash-flow projections. Moreover, the onset of more closely monitored lending, specifically designed credit packages, and negotiated business assessments has been established upon a reorganized structure at regional and local levels.

The reorganization of regional and local bankings has progressively reduced the autonomy of the branch manager. All the banks set a limit on the amount of money a branch manager can lend to one customer; the Midland limit is £300,000. For further borrowing to be sanctioned, the branch manager has to apply to the regional office. A heightened degree of negotiation now takes place between local branch managers and the dealers and merchants providing agricultural inputs. Though the particular design of many of the credit packages linking input sales to credit agencies is developed at the head-office level, the particular negotiation and implementation of these credit packages remains a local affair (see Hawkins 1991). Exclusive agency policy, which links dealers and merchants of agricultural inputs directly to particular agribusiness suppliers, may well be the initial point of contact for the establishment of credit relations. At the local level, facing increasingly fierce competition, the network of relationships that develops between merchants and the local bank branches provides a crucial link in maintaining agro-input markets. Both the bank and the input supplier stand to benefit from the increasing problem of mobilizing working capital faced by the producers. When farmers talk to their local merchants, they are now

[6] The evidence upon which this section draws was collected as part of a joint project between T. K. Marsden and L. G. Soler, INRA (Institut National de le Recherche Agronomique), Paris, 1990. Interviews were conducted with representatives of all the major clearing banks in the United Kingdom and France. The research was funded by the EC and INRA.

invariably also opening a discourse with their more centralized and streamlined bank.

Conclusions

The highly competitive and oliogopolistic market situations facing both the clearing banks and the agro-input suppliers have influenced the growth in leasing companies allied to the agromanufacturers. Banks and agro-input suppliers have joined forces to promote specific short- and medium-term credit packages, providing competitive rates of interest and further tying in the farm sector to their oligopolistic levels of control. Credit relations between farmers, banks, and agribusinesses are thus restructured through increasing the conditions and specifications on which loans are granted. This structure ties the "business-oriented" farmer more directly to the financial/technology "fix." It can enlarge the activities and influence of agribusiness and banks during a period of agricultural disinvestment. Both the control and the ownership of the means of production are further appropriated by financial and industrial capitals. The recession in the farming industry, brought about by a partial reform of European agricultural policy, is thus promoting the restructured financial and agro-industrial sectors to selectively deepen their linkages both with each other and with the farming sector at the international, national, and local levels.

Evidence from France, for instance, also suggests heightened competition among the newly deregulated banking agencies to develop credit arrangements with "upper-range" farmers, based on targeting practices involving computerized financial criteria and assessments. Input supply firms have seen the benefits of cooperation with the restructured banking sector, particularly as market demand for farm inputs has been shrinking (Harrison 1989). Simultaneously, the major clearing banks, now less convinced of the security of agricultural land as a distinctive form of collateral for lending and armed with more immediate and centralized methods of monitoring their branch managers' investment and lending performances, seek to construct more conditional and selective loan arrangements. As Minns (1981) and G. Thompson (1977) imply, banking capital is remarkably adept at reorienting its activities and is now establishing specific new linkages with agro-industry when market closure within a particular sector (such as agriculture) threatens.

In the British context the traditional autonomy of the banking sector vis-à-vis agricultural state intervention, as well as its highly internationalist outlook, distinguish agrofinancial relations from those that have developed in most other countries of Western Europe (Dixon 1991). Nevertheless, the deregulation of the French banking sector, in addition to the harmonization of financial activities across Europe, may suggest a period of structural convergence regarding banking-state relations more generally, and banking and agriculture more specifically (see also Kaye 1989). This evolutionary process is far from smooth or short-run. It may test the more direct support European farmers have received from sector-specific banking structures and state-backed concessionary loans. Though the state—through a reformed CAP—is still likely to intervene through markets or income supports or both, it is likely that it will be the further competitive synergies developed between agro-industrial firms and financial institutions that will more effectively regulate, and increasingly direct, intensive European agriculture.

With regard to the broader processes of restructuring outlined in the first part of this chapter, land-based agriculture becomes increasingly one other sector to which the exploitive synergies of industrial and finance capital are directed. In this regard we can see the resonance of D. S. Glasberg's analysis (1988) of bank hegemony, which stresses the domination of banking in regulating firms, cities, and the nation-state. Indeed, some writers are now interpreting the developments and overall dominance of the banking capital as a restructured form of "fictitious capital," whereby it becomes increasingly mobile and characterized by accelerating speculation and destablization (Green 1990). Nevertheless, the focus on historically specific global processes operating and incorporated within nation-states and particular sectors suggests that banking capital is still dependent on grounded industrial production and consumption, however mobile and influential it has become (Harris 1988). It is the degree of competitive collaboration that banking capital develops with industrial sectors over time and space that seems to be crucial in maintaining its level of hegemony. Moreover, it is the particular combination of relationships and transactions between them that now allows for further penetration into the agricultural sphere at a time of general retrenchment. The question thus is not *whether* banking capital is now directing capital accumulation at global and regional levels, but, more appropriately, *how* the interactions of banking and industrial capital help to shape particular sectors of production and consumption. Bankers are just as crucially concerned with maintaining their procurement of industrial profits as they are with assisting firms to open and reproduce new exploitable markets through their design and allocation of credit instruments. It is, then, this often conjunctural but highly competitive finance-industrial arena that provides a major dynamic in regulating and defining agrarian crisis and development. Historical differences in the position of both sectors through variations in the role of the nation-state can have an important influence in modifying those relations (Vittas 1989).

This chapter has examined at two articulated levels some of the main dynamics shaping the relations between financial capital and industrial capitals. It has done this without necessary recourse to conventional "from-to" evolutionary dualisms, which are now popular in the broader restructuring and regulationist literature (Lash and Urry 1987, Jacques and Hall 1989, Lipietz 1990). Rather, it is regarded as necessary to track through the nature of existing and contingent social and economic relations as they are related to the spheres of production, consumption, and regulation, both at the macro level of the national and international economy and at the micro level of firm or region. So far, the restructuring literature has not sufficiently incorporated analyses of particular sectors (such as the food sector). Similarly, within agrarian sociology, writers are starting to recognize nonagrarian restructuring processes as relevant conditioning factors for assessing and positioning agrarian change. Both of these welcome moves require more than a putative commitment to the recognition of national and historical variation. The study of financial-industrial relations exposes a major sphere in which new rounds of unevenness in production and consumption relations are a principal ingredient, as well as a consequence, of capitalist accumulation in the late twentieth century.

References

Amin, A., and M. Dietrich, 1990. "From Hierarchy to 'Hierarchy': The Dynamics of Contemporary Restructuring in Europe." Paper presented at the Conference of the European Association for Evolutionary Economy, Florence, Italy, November.

Borrie, G. 1987. "Competition, Mergers, and Price Fixing." *Lloyds Bank Review* (April): 1–15.

Cerny, P. G. 1991. "The Limits of Deregulation: Transnational Interpenetration and Policy Change." *European Journal of Political Research* 19: 173–96.

Clark, G. 1989. "Remaking the Map of Corporate Capitalism: The Arbitrage Economy of the 1990s." *Environment and Planning* A 1989: 8, 21, 997–1001.

——. 1990. "Location, Management Strategy, and Workers Pensions." *Environment and Planning* 22: 150–76.

Clauzier, A. 1989. "French Farmers' View of 1992." AMC Review no. 7 (Summer): 8–10.

Coakley, J. 1982. "Finance Capital: A Study of the Latest Phase of Capitalist Development." *Capital and Class* 17: 134–41.

——. 1988. "International Dimensions of the Stock Market Crash." *Capital and Class* 3:16–21.

Cooke, P. 1989. "Critical Cosmopolitanism: Urban and Regional Studies into the 1990s." *Geoforum* 20(2): 241–52.

Cumming, C. 1987. "The economics of securitisation." *FRBNY Quarterly Review* 12(3): 10–22.

Dickson, T. 1991. "The Spirit of 1992." *Financial Times*. Special Issue (November): 2.

Dixon, M. 1991. *Banking in Europe*. London: Wiley.

Dymski, G. A. 1990. "Money and Credit in Radical Political Economy: A Survey of Contemporary Perspectives." *Review of Radical Political Economics* 22: 38–65.

Evans, T. 1989. "Dollar Is Likely to Rise, Fall, or Stay Steady, Experts Agree." *Capital and Class* 13: 10–15.

Glasberg, D. S. 1988. *The Power of Collective Purse Strings: The Effect of Bank Hegemony on Corporations and the State*. San Francisco: University of California Press.

Glyn, A. 1988. "The Crash and Real Capital Accummulation." *Capital and Class* 3: 21–24.

Goodman, D., and M. Redclift, eds. 1989. *The International Farm Crisis*. London: Macmillan.

Green, G. 1988. *Finance Capital and Uneven Development*. Boulder, Colo.: Westview Press.

——. 1990. "The International Division of Labour and Fictitious Capital." Paper presented at the annual meeting of the American Sociological Association, Washington D.C., August.

Harris, L., J. Coakley, M. Cransdale, and T. Evans. 1988. *New Perspectives on Financial Systems*. London: Croom Helm.

Harrison, A. 1989. "The Changing Financial Structure of Farming." Working paper no. 13. University of Reading: Centre for Agricultural Strategy.

Harvey, D. 1989. *The Condition of Postmodernity*. Oxford: Basil Blackwell.

Hawkins, E. 1991. "Changing Technologies: Negotiating Autonomy on Cheshire Farms." Ph.D. diss., South Bank Polytechnic, London.

Hirst, P., and J. Zeitlin. 1989. "Flexible Specialisation and the Competitive Failure of U.K. Manufacturing." *Political Quarterly* 60: 164–78.

——. 1991. "Flexible Specialisation versus Post-Fordism: Theory, Evidence, and Policy Implications." *Economy and Society* 20(2): 1–56.

Jacques, M., and S. Hall, eds. 1989. *New Times*. London: Lawrence & Wishart.

Jessop, B. 1989. "Regulation Theory, Post-Fordism and the State: More Than a Reply to Werner Bonefield." *Capital and Class* 4: 147–67.

——. 1990. "Regulation Theories in Retrospect and Prospect." *Economy and Society* 19(2): 153–216.

Kaye, J. 1989. "The Implications of 1992 for U.K. Agricultural Finance: A Practical Banker's View." Paper presented at the meeting of the Agricultural Economics Society, Manchester, U.K., December.

Keat, R. 1990. "Starship Britain or Universal Enterprise?" In *Enterprise Culture*, ed. R. Keat and N. Abercrombie. London: RKP.

Lash, S., and J. Urry. 1987. *The End of Organised Capitalism.* Cambridge: Polity.

Lipietz, A. 1990. "The Debt Problem, European Integration, and the New Phase of World Crisis." *New Left Review* 26: 37–50.

Lomax, D. 1989. "The 'Big-Bang' 18 Months After." *Barclays Bank Review* 4(2): 18–31.

McMichael, P., and D. Myhre. 1991. "Global Regulation vs. the Nation-State: Agro-Food Systems and the New Politics of Capital." *Capital and Class* 43: 83–105.

Marsden, T. K., A. Flynn, and N. Ward. 1991. "Managing Food: A Critical Analysis of the British Experience." Paper presented at the Globalization of Agriculture and Food Conference, University of Missouri, Columbia, June.

Marsden, T. K., and J. K. Little, eds. 1990. *Political, Social, and Economic Perspectives on the International Food System.* Avebury, U.K.: Gower.

Marsden, T. K., P. Lowe, and S. Whatmore. 1990. *Rural Restructuring: Global Processes and Their Responses.* Vol. 1 of *Critical Perspectives on Rural Change.* London: Fulton.

Marsden, T. K., and J. Murdoch. 1990. "Restructuring Rurality: Key Areas for Development in Assessing Rural Change." Working Paper no. 4. Economic and Social Research Council Countryside Change Initiative, University of Newcastle.

Marsden, T. K., S. Whatmore, and R. J. C. Munton. 1990. "The Role of Banking Capital and Credit Relations in British Food Production. In *Political, Social, and Economic Perspectives on the International Food System,* ed. T. K. Marsden and J. K. Little. Aldershot, U.K.: Gower.

Martin, R. 1989. "The Reorganisation of Regional Theory: Alternative Perspectives on the Changing Capitalist Space Economy." *Geoforum* 20(2): 187–201.

Miller, P., and N. Rose. 1990. "Governing Economic Life." *Economy and Society* 19(1): 1–31.

Minns, R. 1981. "A Comment on Finance Capital and the Crisis in Britain." *Capital and Class* 14: 98–110.

Reutan, G. 1989. "The Money Expression of Value and the Credit System: A Value-Form Theoretic Outline." *Capital and Class* 4: 121–41.

Rowthorn, R. 1971. "Imperialism in the Seventies." *New Left Review* (September/October): 15–32.

Rybczynski, T. M. 1989a. "Corporate Restructuring." *National Westminster Bank Review* (August): 18–28.

——. 1989b. "Financial Systems and Industrial Restructuring." *National Westminster Bank Review* (May): 3–13.

Scouller, J. 1989. "The United Kingdom Merger Boom in Perspective." *National Westminster Bank Review* (May): 14–31.

Terry, N. 1985: "The 'Big Bang' at the Stock Exchange." *Lloyds Bank Review* (April): 16–30.

Thompson, G. 1977. "The Relationship between the Financial and the Industrial Sector in the United Kingdom Economy." *Economy and Society* 6(3): 235–83.

Thrift, N., and A. Leyshon. 1988. "The Gambling Propensity: Banks, Developing Country Debt Exposures, and the New International Financial System." *Geoforum* 19: 55–69.

Tsoukalis, L., ed. 1985. *The Political Economy of International Money: In Search of a New Order.* London: Sage Publications.

Vittas, D. 1987. "Banks' Relations with Industry: An International Survey." *National Westminster Bank Review* (May): 2–13.

White, B., and D. Vittas. 1986. "Barriers to International Banking." *Lloyds Bank Review* (July): 19–31.

Williamson, P., and D. F. Lomax. 1989. "Lessons for Banking from the 1980s and the Recent Past." *National Westminster Bank Review* (May): 2–15.

5

Industrial and Labor Market Transformation in the U.S. Meatpacking Industry

Kathleen Stanley

Since the late 1970s the U.S. Midwestern meatpacking industry has been marked by increased levels of competition and concentration, plant closings and relocations, and decreases in the overall number of workers and their wages. Accompanying these changes has been a thorough transformation of the labor force. Most noticeable has been the dramatic increase of foreign-born workers—including both immigrants and refugees—in the industry. U.S.-born workers have also migrated to these jobs, often from other rural areas, as have growing numbers of women.

The transformation of the meatpacking industry and its labor force must be viewed within a context of both industrial restructuring and rural crisis. In recent years increased foreign competition has led to downward pressures on the historically high wages of U.S. workers, leading many industries, including meatpacking, to relocate to relatively low-wage areas (Bluestone and Harrison 1982, Harrison and Bluestone 1988). Rural communities welcomed these industries as the farm crisis weakened their economies and sent significant numbers of family farmers in search of wage labor to replace or supplement farm income. Though industrial restructuring in meatpacking has provided

This research was funded by the Division of Immigration Policy and Research, Bureau of International Labor Affairs, U.S. Department of Labor, and by a Rural Policy Fellowship from the Woodrow Wilson Foundation. I thank Rita Argiros, Bob Bach, Dean Braa, Fred Buttel, Phil McMichael, Martin Murray, Demetrios Papademetriou, and Dale Tomich for their comments and input on earlier drafts of this essay.

much-needed rural employment, it has also generated uneasiness over the industry's growing concentration and the potential effects this concentration may have on livestock producers and workers in rural communities.

Changes in the meatpacking industry illustrate more general trends in food processing. The globalization of the food system, health concerns, and changing lifestyles have contributed to heightened competition for the "food dollar." Increased competition and the relative inelasticity of food markets have led to an emphasis on value-added food processing as a way to boost profits. The rise in women's labor force participation in the United States is one factor that has led to the widespread use of highly processed foods at home and the proliferation of fast-food restaurants. Even fresh foods such as meats, poultry, and vegetables arrive at grocery stores in new forms: cut up, boned, precooked, and so on. As a result, food-processing jobs are concentrated more and more in the manufacturing sector. Both the nature and location of food-processing jobs have changed. As production moves from one location to another, workers are displaced. Many sectors of the food-processing industry—from meatpacking to restaurants—are relying more heavily on nonunionized immigrant labor (Papademetriou et al. 1989).

In general labor market theory, the volume of immigration and the distribution of foreign-born workers should be explained by imbalances in labor supply and demand. Since the 1960s, however, the rate of immigration has increased despite high rates of unemployment. In the specific case of meatpacking, foreign-born workers have entered the industry during a period of widespread layoffs and plant closures. Rather than serving as a supplementary labor supply in economic sectors characterized by labor shortages, immigrants have concentrated in specific labor markets, such as meatpacking, that have a special demand for these workers (Portes and Bach 1985). Dual labor market theorists (e.g., Piore 1979) have argued that this special demand is linked to structural divisions among firms that are manifest in different levels of job security, wage rates, and opportunities for mobility. Others have argued that immigrants are of special value because they are easily exploited. Still others have argued that the demand for immigrant labor may be due less to the need for low-wage labor than to the flexibility that new supplies of labor create during periods of industrial restructuring (Sassen-Koob 1985, Morales 1984).

The research reported here is a case study of the meatpacking industry and is based in part on fieldwork conducted from 1988 to 1990 in several communities. These communities were chosen to highlight key variables in the restructuring of the meatpacking industry. Data were collected from published sources and through interviews with key informers (industry officials, social service providers, community leaders, union officials, and journalists). Increasingly, researchers have moved to local labor market case studies to uncover the complex interrelationship of social and economic factors that explains the demand for new types of labor in areas of economic decline or restructuring. A case-study approach makes possible an analysis of the multiple circumstances in which labor recruitment is organized and the impact that the presence of immigrant workers has on rural labor markets.

Historical Overview

Before the Civil War meatpacking was seasonal and localized. The work force was composed primarily of area farmers who supplemented their incomes in slaughterhouses during normally idle winter months. Skilled butchers found year-round employment in the retail fresh meat trade (Skaggs 1986:108).

Modern meatpacking emerged in the last decades of the nineteenth century and was dependent on the development of refrigerated railroad cars for the preservation of meat while in transit to market. Large meatpacking companies quickly developed in cities that were close to livestock markets and transportation facilities. Chicago was especially important as the center of the new meatpacking industry, although smaller Midwestern cities such as Kansas City, Omaha, St. Louis, and Sioux City were also important.

With the transformation of the industry from seasonal to year-round employment, temporary workers were replaced by full-time, primarily unskilled laborers. Employment in the industry expanded rapidly, from only 8,366 in 1870 to over 60,000 by the turn of the century (Skaggs 1986:108). Unskilled positions were filled more and more by immigrants. By 1910 only 19 percent of the labor force was white and native-born; nearly half the work force was Eastern European (Barrett 1984:40).

Black migrants from the rural South also began to enter the industry in the 1910s. Their numbers grew especially during World War I, as European immigration slowed and wartime industries expanded. Meatpacking was one of the primary industrial employers of African Americans during this period, often accounting for half the manufacturing employment of black men. By 1920 African Americans constituted 15 percent of all employment in meatpacking and 21 percent of unskilled workers. Many African Americans initially entered meatpacking as strikebreakers (Herbst 1932). In fact, meatpacking plant owners opposed policies advocated by organized labor and civic groups that would have slowed the migration of Southern blacks to the North and Midwest, and some employers used labor recruiters to encourage black migration—though after the chain of migration began there was little need for direct recruitment (Fogel 1970:26). During the Depression, black employment in the industry declined as the demand for meat sagged and as meatpacking jobs became more attractive to white workers. Black internal migrants were joined by Mexican immigrants traveling up the Midwestern route from the Central Valley in Texas to Chicago. By 1968 approximately 4 percent of the meatpacking work force was composed of Hispanics (mostly Mexican immigrants), and another 11 percent was composed of African Americans. Total minority employment was around 15.5 percent (Fogel 1970:76).

Changes in the industry during the interwar period had a significant impact on the demand for labor. Forced by the federal government to break up large, vertically integrated companies and to refrain from price fixing, the industry was pushed by a new wave of competition toward greater mechanization and the search for labor-saving techniques. Semiskilled jobs replaced the unskilled, manual labor of earlier workers. The technological innovation of this period should not be overemphasized, however; "basic plant design had changed very little since the 1880s, most facilities remaining multistoried structures that relied on animal power to move the raw material to the top floors, where the beasts were killed, and gravity to propel the chains that carted carcasses along disassembly lines" (Skaggs 1984:154). When the absolute demand for labor again rose during World War II, the industry employed ever-increasing proportions of black and female workers. By the 1950s over 20 percent of meat-industry workers were women, a figure that reflects the higher demand for female workers during World War II (Fogel 1970:53).

Since World War II the industry has been undergoing almost continuous restructuring. The development of the interstate highway system freed meatpackers from dependence on the railroads and allowed them to move away from plants situated at railheads in urban centers. Simultaneously, "a second technological revolution hit meatpacking. New methods and machinery began to appear: stunners, mechanical knives and hide skinners, power saws, electronic slicing and weighing de-

Table 5.1. Decline in meatpacking wages

	No. of production workers	Average hourly wages (US$)		Meatpacking wage as percentage of manufacturing wage
		Meatpacking	Manufacturing	
1969	143,500	$3.66	$3.19	115
1971	145,600	4.17	3.57	117
1973	137,600	4.71	4.07	116
1975	134,500	5.61	4.81	117
1977	137,000	6.44	5.63	114
1979	134,300	7.73	6.69	116
1981	129,200	8.98	7.98	113
1983	115,400	8.58	8.84	97
1985	123,300	8.10	9.52	85
1987	115,300	8.37	9.91	84
1989	121,100	8.63	10.47	82

Sources: Bureau of Labor Statistics, *Employment and Earnings:* vol. 16, no. 9 (March 1970); vol. 18, no. 9 (March 1972); vol. 20, no. 9 (March 1974); vol. 22, no. 9 (March 1976); vol. 25, no. 3 (March 1978); vol. 27, no. 3 (March 1980); vol. 29, no. 3 (March 1982); vol. 31, no. 3 (March 1984); vol. 33, no. 3 (March 1986); vol. 35, no. 3 (March 1988); vol. 37, no. 3 (March 1990).

vices. . . . Labor production rose by 15 percent from 1954 to 1958 . . . [and] employment began to fall" (Brody 1964:241–42). Beginning in the 1950s obsolete, multistoried plants were closed in traditional urban strongholds as the industry relocated to rural areas near feedlots where land prices were lower. Many of the new plants were located in "right-to-work" states, where the price of labor was also significantly lower. But collective bargaining in the industry had been established at the end of World War II, when many meatpacking firms had signed master agreements with the unions (the Amalgamated Meat Cutters, Butcher Workmen of North America, and the Packing House Workers Organizing Committee). In 1969 wages in meatpacking were about 15 percent higher than in manufacturing in general (Table 5.1).

By the 1960s northern Iowa and southern Minnesota had become the geographic center of the industry. Owners could hire only a small fraction of the labor force from the local community and had to seek new supplies of workers, which recomposed the labor force. African Americans accounted for 24 percent of the meatpacking labor force in the older urban plants. Although some were relocated with the plants under collective-bargaining agreements, by the late 1960s they made up only about 2 percent of the meatpacking work force in Iowa and Minnesota (Fogel 1970:77). The new supplies of labor included rural whites and an expanding number of Mexican Americans and Mexican immigrants. A relatively small number of workers were women—only 14 percent in 1967 (Fogel 1967:14).

Recent Restructuring in the Meatpacking Industry

The most recent stage of restructuring in the industry, which began in the late 1970s, has been particularly turbulent. The major features of this restructuring include changes in the production process, a new wave of plant relocations, intensified competition, and a significant reduction of wage scales. These changes have necessitated new labor supplies, which have been filled largely by immigrants, refugees, and U.S.-born workers who have migrated to jobs in this industry.

New marketing strategies have had an important impact on production. Foremost among these innovations was the advent of "boxed beef" and later "boxed pork." Previously, carcasses were sent to retail stores as "hanging halves," which required considerable additional processing by in-store butchers. Now they are processed at the packing plant into "primal cuts," which are vacuum-packed in boxes and shipped to supermarkets.[1] Supermarket employees "have to do little more than open the boxes, slice up the chunks almost like loaves of bread, and place the retail cuts on the meat counter" (Swanson and Schultz 1982:181). Supermarkets save on butchering costs because they no longer require skilled butchers, and they have less waste in the form of bones and fat.

Production in the meatpacking industry also has become more automated. The new one-story plants rely on mechanized disassembly lines instead of the gravity-driven lines of the older, multistoried plants. The combination of automation and a greatly expanded role of packing plants in processing meat has resulted in a more detailed division of labor. New production methods have produced a further reduction of skilled meatpacking jobs while simultaneously raising productivity by giving management greater control over the pace of production. In spite of increased mechanization, most jobs in the processing (fabrication) departments of meatpacking plants are manual, involving hand-held knives. Consequently, increased productivity has been largely the result

[1] As an extension of this trend, some meatpackers now offer branded beef—individual cuts, such as steaks, which bear the meatpacker's label. Unlike beef sold in oxygen-permeable films (which allow the beef to turn bright red), individually vacuum-packed cuts retain their natural bluish or purple-red color. Though this method extends their shelf-life up to fifteen days, it has not caught on with consumers, who depend on meat color as an indicator of freshness.

of a finer division of tasks (facilitating rapid repetition) and increased chain speeds (see also Skaggs 1984:191–92).

Accompanying these changes in production has been another wave of plant relocation. New plants have been built in sparsely populated rural areas, near the huge feedlots that emerged on the High Plains (Nebraska, Kansas, Colorado, and Texas) as irrigation agriculture increased the supply of feed grains in the post–World War II era. Relocation brought significant savings by eliminating transportation costs associated with the shipment of live animals. Other advantages to relocation included cheap land for the sprawling single-story plants and a variety of financial incentives provided by rural communities eager to attract new industries. Geographic relocation has allowed firms to take advantage of rural wage scales and to avoid the strong labor unions that had developed in the industry. Although relocation has brought with it certain benefits for the meatpacking firms, it has also confronted them with the challenge of attracting and retaining an adequate labor force for large plants in small rural towns.

Competition in meatpacking was fierce during the 1980s. New aggressive firms in the industry were able to both increase productivity and reduce wages.[2] Although the demand for red meat had fallen by about 20 percent since the mid-1970s and foreign competition had increased, these firms had more than doubled their share of the market (Weingarten 1988, Massey 1988).[3] Other firms were forced to either close or compete on new terms. The result was a massive shake-up in the industry, with numerous plant closings and wage reductions across the board. A key determinant of competitiveness was the cost of labor. In addition to relocating, many firms attempted to escape union contracts and higher labor costs by shutting down and reopening at substantially lower wage rates or by filing for reorganization under Chapter 11 bankruptcy law proceedings (see also Moody 1988, Hage and Klauda 1989).

Employment in meatpacking was thus becoming less and less attractive. Between the mid-1970s and the early 1980s wages were slashed by as much as 50 percent, and an emphasis on productivity resulted in line speed-ups and increasing injury rates. When compared with average

[2] The three largest meatpacking firms are ConAgra (which purchased both Monfort and Swift in the mid-1980s), IBP (owned by Occidental Petroleum), and Excel (owned by Cargill). Together they control approximately three-quarters of the fresh meat market in the United States.

[3] As domestic markets have contracted, meatpackers have tried to expand their overseas markets, especially in Japan and other Pacific Rim countries. Trade agreements with Japan in the late 1980s, for example, have expanded the volume of U.S. beef imports.

manufacturing wages, wages in meatpacking fell by nearly 30 percent between 1979 and 1989 (see Table 5.1). Moreover, meatpacking is one of the nation's most dangerous industries during the 1980s: about one-third of its workers were injured each year, and in Iowa the injury rate reached 43 out of 100 workers (Pezaro et al. 1985, Fowler 1988). The largest fines in the history of the Department of Labor's Occupational Safety and Health Administration (OSHA) were levied in 1988 against two meatpacking firms for covering up the extent of occupational injuries and for failing to institute programs to reduce carpal tunnel syndrome, a crippling wrist disorder common among meatpacking workers.

Not surprisingly, labor turnover is quite high in the industry. Estimates range from around 50 to over 100 percent annually, and there is considerable variation among firms. Turnover rates are even greater when new plants first open—as high as 150 to 500 percent (Ackland 1983a, 1983b). It appears that most plants eventually develop a bifurcated labor force—a stable core of workers and another segment that turns over almost constantly. As a result, the industry requires a steady stream of new workers.

The Transformation of the Labor Force in Meatpacking

Distribution of Foreign-Born Workers

Accompanying the restructuring of the meatpacking industry has been a dramatic transformation of its labor force. By all accounts, the number of immigrant and refugee workers, the number of native-born migrant workers, and the proportion of women workers have increased since the late 1970s.[4] As indicated by Table 5.2, the composition of the labor force within the industry varies considerably from plant to plant.

The six plants in Table 5.2 were part of a survey of the three largest meatpacking firms (ConAgra, IBP, and Excel) conducted in the summer of 1988. At that time, these firms encompassed approximately thirty plants and 35,000 employees. The firms were asked to provide information on total employment, the composition of the labor force by race and sex, turnover rates, and wage rates. Data were obtained for a total of fifteen plants. Although the response to the survey was highly variable, the data collected, in combination with information from other sources, do permit some limited generalizations.

[4] Although meatpacking historically employed many immigrant workers, in the immediate post–World War II era the labor force was predominantly white and native-born.

Table 5.2. Composition of the labor force in selected meatpacking plants

	Plant 1	Plant 2	Plant 3	Plant 4	Plant 5	Plant 6
Total no. of employees	1,800	1,700	780	640	680	800
% Female	22	29	24	25	13	13
% White	39	39	37	97	92	63
% Black	4	1	33	<1	<1	14
% Hispanic	38	59	13	1	<1	9
% Asian	18	1	17	2	7	14

Notes: Plant 1 is a newer beefpacking plant in a rural area of the High Plains. The town has a population of less than 20,000. Plant 2 is a beefpacking plant in an area with a large Hispanic population. The town has a population of over 50,000. Plant 3 is a beef fabrication plant (no slaughter) in a city with a population of over 250,000. Plant 4 is a porkpacking plant in a town with a population of about 25,000. Plant 5 is a porkpacking plant in a town with a population of 10,000. Plant 6 is a porkpacking plant in a city with a population of about 200,000.

Foreign-born workers, for example, are most often found in the largest plants and firms (see also Griffith 1990). Black employment in the industry is generally confined to plants in larger cities and is declining as these plants close.[5] The number of women working in meatpacking is increasing continuously. In some plants women now make up close to half of the labor force—although the average is closer to 20 to 25 percent.

When the initial survey was conducted in 1988 the beef and pork sectors of the industry exhibited quite different labor market patterns. In the porkpacking plants of the Corn Belt states (especially Iowa and Minnesota), the majority of the labor force was white and native-born (often upward of 90 percent), especially in the rural pork plants. Urban pork plants were characterized by much greater diversity. Until the late 1980s, most foreign-born workers in the pork sector were Southeast Asian refugees, and there were relatively few of them.

Immigrant and refugee workers were most heavily concentrated in the beefpacking plants of the High Plains. In these plants, Hispanics (many of whom were Mexican immigrants) and Asians (virtually all of whom were officially recognized refugees) typically made up half or more of the labor force. Of this number, the majority were Hispanic. In some beef plants—generally those located in areas with large Hispanic populations—Hispanics alone made up more than half of the labor

[5] For example, Plant 3 (from Table 5.2), an urban plant with a large proportion of African American workers, has closed since the initial survey was conducted.

force. Asians made up anywhere from 2 to 25 percent, but typically 10 to 15 percent, of a plant's labor force. Interviews with informants indicated that the remainder of the labor force in these plants was composed of U.S.-born workers who migrated to the plants, often, though by no means exclusively, from other rural areas.

These variations remain evident in the early 1990s and are the result of differences in size, location, and ownership. The largest of the beef plants frequently employ over 1,500 workers. Many beef plants have relocated to towns with populations of 20,000 or less, which makes the acquisition of a local work force virtually impossible. The operation of these plants thus requires massive influxes of workers, especially when high turnover rates are taken into consideration. In contrast the largest pork plants typically employ 600 to 800 workers. Moreover, they tend to be located in areas with greater population densities and more owners of small farms (who are more likely to seek off-farm employment).[6]

The restructuring of the industry began in the beef sector, but by the late 1980s, spread to porkpacking. As restructuring of the pork sector proceeds, the extent of these differences in the ethnic composition of the labor force will be minimized over time. When additional fieldwork was conducted in the summer of 1990, informants in towns with porkpacking plants indicated that they were currently experiencing large influxes of Mexican workers and attributed this to the entry into pork of newer firms that, as indicated above, have very different labor relations strategies and, as a consequence, very different labor supply needs. Most informants (including spokespersons from the meatpacking firms) agreed that the local labor supply in some of these areas had been exhausted. The beef- and porkpacking sectors of the meatpacking industry are thus increasingly similar in structure.

Recruitment of Workers

The same few large corporations now control major portions of the industry in both the beef and pork sectors and actively recruit foreign-born workers. Some of the foreign-born, although by no means the majority, entered the industry initially as replacement workers during strikes. Asian refugees, especially, were arriving in large numbers just

[6]The Corn Belt states were especially hard hit during the farm crisis of the early 1980s, which added to the pool of potential meatpacking workers. Newspaper reports at the time indicated that many replacement workers during the 1985–86 Hormel strike in Austin, Minnesota, were struggling or displaced farmers from the region (Serrin 1986, see also Hage and Klauda 1989).

as the industry shake-up was beginning. Some of the largest companies employ a staff of labor recruiters who target areas with high unemployment. The recruiters travel to these areas, conduct interviews, and offer jobs to those willing to migrate. Workers who move more than one hundred miles receive 16 cents per mile in compensation and advances of up to $200 to help them get settled.[7] In 1990 informants indicated that most new recruits were arriving in the Midwest from southern Texas and northern Mexico.

Many workers find meatpacking jobs through state and local employment agencies, and the structure of federally supported job-placement programs has had an important impact on the recomposition of the labor force in the industry. Programs designed to promote self-sufficiency among refugees and to move seasonal agricultural workers into permanent employment have encouraged these individuals to take meatpacking jobs. The service providers interviewed for this project between 1988 and 1990 universally mentioned the need to find immediate, full-time employment for their clients in order to comply with the guidelines of their federal funding. Because in the Midwest meatpacking jobs are plentiful and open to non-English-speaking workers, the service providers must often encourage their clients to apply for these jobs even when they themselves have reservations, which many do, about the working conditions in the plants.

Other federal programs also encourage meatpackers to hire certain categories of workers. According to one state employment official in Kansas, the meatpacking firms make "extensive" use of two employment programs that result in significant subsidies to the industry. Under the Job Training Partnership Act, federal funds will pay for half the basic wages of disadvantaged workers during the initial training period. In rural areas, this program is frequently used to help move seasonal agricultural workers into full-time employment in meatpacking. Another important program is the Targeted Jobs Tax Credit program designed to help disadvantaged workers escape chronic unemployment. Under this program employers can claim tax credits equal to 40 percent of the first $6,000 of qualified wages paid to the applicant during the first year on the job. At one plant in Iowa up to 25 percent of new workers qualify under this program.[8]

Recruiting also takes place through the social networks of the workers themselves. Workers at some plants receive bonuses for recruiting

[7] Interview with Corporate Director of Staffing (Summer 1990).
[8] Interview with local Iowa Job Services Director (Summer 1990).

family members and friends. A typical bonus is $150 if the new recruit makes it past the probationary period—usually 90 or 120 days. The formation of ethnic communities in meatpacking towns encourages further migration as older residents frequently act as formal sponsors (in the case of refugee resettlement) or assist informally in locating housing and services. In fact, the lack of African American communities of any size in most rural Midwestern towns is often seen as the major reason why so few blacks are willing to migrate to these jobs.[9]

Foreign-born workers are attracted to meatpacking jobs because neither English language proficiency nor previous experience is required and because these jobs pay quite well by regional standards. The alternative to meatpacking employment for these workers is usually custodial or agricultural work at minimum or subminimum wages; starting wages for meatpacking are around $6 to $6.50 per hour. In the Midwest there are few other job opportunities, especially in manufacturing. For Mexican migrants meatpacking frequently provides a year-round or seasonal alternative to agricultural work. Refugee workers appear to be particularly dependent on meatpacking employment. In the Midwest, 62 percent of Southeast Asian refugees employed in manufacturing are found in meatpacking.[10]

Women enter the industry for many of the same reasons as foreign-born men: it requires neither experience nor English (for those who are foreign-born themselves) and pays considerably more than most so-called women's jobs. Mechanization and further processing of carcasses into boxed beef and pork have also made meatpacking jobs more accessible to women. Meatpackers clearly realize that women, especially those with young children, constitute a pool of potential labor that has as yet been imperfectly tapped. Efforts to attract more women workers are reflected in promises of daycare facilities at new packing plants. It is not uncommon for married couples to be employed at the same packing plant and to coordinate their work schedules to accommodate childcare responsibilities.

[9] Though this is a widespread perception it does not provide a satisfactory answer. Asians and Hispanics (with some exceptions) are also newcomers to these towns. The presence or absence of ethnic communities is not sufficient to explain these differences, which also depend on access to services and previous work migration patterns.

[10] *Annual Survey of Southeast Asian Refugees*. 1987. Office of Refugee Resettlement, United States Department of Health and Human Services.

Impact of Immigrant and Refugee Employment

Industrial restructuring in the meatpacking industry has had wide-ranging effects throughout the Midwest. Although a full treatment of these effects is not possible here, the process of labor force recomposition, which has had an important impact, should be discussed. Not surprisingly, this process has had very different outcomes for the industry, workers, unions, and rural communities.

The effect on the industry has been twofold. On one hand, the need for new supplies of labor emerged from dynamics within the industry. At the same time, however, the process of restructuring was itself dependent on the availability of large numbers of new workers. Plant closures and relocations, lower wages, and increased injury rates have led to the decline of the traditional labor force, even as higher turnover rates have resulted in a higher absolute demand for labor. This demand has been met by Asian refugees, Mexican immigrants, and native-born white and Mexican-American migrants, many of them actively recruited by the meatpacking firms. Although their employment in the industry has coincided with the deterioration of wages and working conditions, their employment has been more the result of that deterioration, rather than its cause.

Until the late 1970s, meatpacking was a high-wage industry that provided workers with middle-class incomes. It was not uncommon for workers who had entered the plants as young adults to retire many years later with good retirement pensions (see, for example, Hage and Klauda 1989). Meatpacking jobs no longer allow for such social mobility. For workers with few other job opportunities though, such as women and the foreign-born, meatpacking has provided relatively high-paying, if dangerous, jobs. Some workers are promoted to more highly paid supervisory positions—most firms now seek out bilingual supervisors as a way of attracting and retaining non-English-speaking workers—but such positions are relatively few. As the higher turnover rates indicate, meatpacking jobs are temporary for most workers. According to informants, even those who form the stable core of the labor force rarely last more than three to five years. By all accounts, the labor force is overwhelmingly young, with most workers in their early twenties. Given the physically demanding nature of the work, very few of these workers will be able to continue in the plants as they get older, even if they escape serious injury in the intervening years.

A key effect of the industry's restructuring has been the decline of union strength. The relocation of packing plants allowed firms to escape or avoid union contracts, and organizing efforts in the new plants have met with little success because of high employee turnover. The presence of large numbers of foreign-born workers has also hindered organizing attempts. Most workers entered the industry as union strength was declining and in areas where unions were nonexistent. The unions are often viewed as ineffective, since they have not been able to protect workers from wage give-backs and line speed-ups. In an effort to recruit foreign-born members, the unions now employ bilingual organizers. There were a few successful organizing drives in the early 1990s, but these were largely confined to new plants in old packing towns with strong union traditions. Key segments of the industry, including many of the largest plants, remain unorganized.

The restructuring of the meatpacking industry has resulted in economic development and much-needed employment in rural communities. It has also created new social problems. As plants first open, homelessness in these communities becomes an immediate problem. Large numbers of new workers create housing shortages, and most workers arrive with very little money. Many end up living temporarily in cars or tents in city and state parks, dependent on local food pantries for their subsistence. Local school systems must mobilize quickly to meet the needs of non-English-speaking students in what often had been ethnically homogeneous communities. Because many workers do not receive medical benefits for six months, local health care delivery systems are strained. Over time, as communities organize to meet these challenges problems become less acute. In spite of the costs associated with this form of economic development, many community members come to appreciate and take pride in their new ethnic diversity (Stull et al. 1990).

Conclusions

Since the late 1970s, the labor force in the meatpacking industry has been dramatically transformed. A major component of that transformation has been an increase in the number of immigrant and refugee workers. It has been argued here that the distribution of foreign-born workers in the industry results from the interplay of several factors: the nature of restructuring, the locally available labor supply, and the nature of programs designed to move immigrants and refugees into full-time employment.

The recomposition of the labor force in meatpacking has been an integral part of the process of industrial restructuring. Pressed by both rising competition and a falling market for red meat, meatpacking firms have turned to immigrant, refugee, native-born migrant, and women workers as a way of controlling labor costs. Lower wages, poor working conditions, and increased injuries have led to high rates of labor turnover, which have made a large flow of new workers necessary.

It is important to note, however, that in the case of meatpacking the demand for immigrant labor was not the result of an absolute shortage of workers. Rather, this demand was created by the industry as firms pursued restructuring strategies that relocated jobs and made them less attractive to the established labor force. Even though meatpacking pays foreign-born workers relatively high wages, the meatpacking firms must actively recruit new workers and tap into the social networks of immigrant groups in order to acquire an adequate labor supply. The dynamics of labor market reorganization involve a complex interaction of social and economic factors that go well beyond traditional explanations focusing on supply and demand.

The entry of large numbers of foreign-born workers into meatpacking (as well as agriculture and other food-processing industries) highlights the nature of globalization in the agro-industrial food system. Globalization involves not just the flow of commodities across nation-state boundaries, but also the international movement of labor. The fluidity of capital, labor, and commodities throughout the world economy multiplies the number of restructuring strategies that can be successfully pursued by capitalists as they attempt to overcome local or regional limits to accumulation. Workers throughout the world economy now find themselves in competition with one another, although this competition is often spatially diffuse. The internationalization of capital and, increasingly, of labor throughout the system have allowed capitalists to reduce substantially their labor costs and to demand concessions from communities and nations in exchange for economic "development." As the case of meatpacking demonstrates, processes of international capital accumulation have very real and observable consequences for rural workers and communities in the United States.

References

Ackland, Len. 1983a. "Recession and 'New-Breed' Meatpackers Buffet Industry." *Chicago Tribune,* June 5: sec. A, pp. 1, 8.

——. 1983b. "Along with Jobs, the Packers Bring Share of Problems." *Chicago Tribune,* June 8: sec. A, pp. 1, 8.

Barrett, James R. 1984. "Unity and Fragmentation: Class, Race, and Ethnicity on Chicago's South Side, 1900–1922." *Journal of Social History* 18 (Fall): 37–55.

Bluestone, Barry, and Bennett Harrison. 1982. *The Deindustrialization of America.* New York: Basic Books.

Brody, David. 1964. *The Butcher Workmen.* Cambridge: Harvard University Press.

Fogel, Walter A. 1970. *The Negro in the Meat Industry.* The Racial Policies of American Industry, Report no. 12. Philadelphia: University of Pennsylvania, Wharton School of Finance and Commerce.

Fowler, Veronica. 1988. "Iowa Meat-Packing Injuries Well Above National Average." *Des Moines Register,* March 20: sec. B, pp. 1, 6.

Griffith, David. 1990. "The Impact of the Immigration Reform and Control Act's (IRCA) Employer Sanctions on the U.S. Meat and Poultry Processing Industries." Unpublished manuscript.

Hage, Dave, and Paul Klaude. 1989. *No Retreat, No Surrender: Labor's War at Hormel.* New York: William Morrow and Company.

Harrison, Bennett, and Barry Bluestone. 1988. *The Great U-Turn: Corporate Restructuring and the Polarizing of America.* New York: Basic Books.

Herbst, Alma. 1932. *The Negro in the Slaughtering and Meatpacking Industry in Chicago.* Reprint. New York: Arno and the The New York Times, 1971.

Massey, Barry. 1988. "Meatpacking Industry Is Concentrated." *Garden City Telegram,* May 27.

Moody, Kim. 1988. *An Injury to All: The Decline of American Unionism.* New York: Verso.

Morales, Rebecca. 1984. "Transitional Labor: Undocumented Workers in the Los Angeles Automobile Industry." *International Migration Review* 17(4): 570–96.

Papademetriou, Demetrios G., Robert L. Bach, Kyle Johnson, Roger G. Kramer, Briant Lindsay Lowell, and Shirley J. Smith. 1989. *The Effects of Immigration on the U.S. Economy and Labor Market.* Washington, D.C.: U.S. Department of Labor, Bureau of International Labor Affairs.

Pezaro, Alan, Stanford Leffingwell, and Kathryn R. Mahaffey. 1985. "Occupational Injuries in the Meatpacking Industry, United States, 1976–1981." *Morbidity and Mortality Weekly Report; CDC Surveillance Summaries* 34 (1SS): 29SS–32SS.

Piore, Michael. 1979. *Birds of Passage: Migrant Laborers in Industrial Societies.* New York: Basic Books.

Portes, Alejandro, and Robert L. Bach. 1986. *Latin Journey: Cuban and Mexican Immigrants in the United States.* Berkeley: University of California Press.

Sassen-Koob, Saskia. 1985. "Changing Composition and Labor Market Location of Hispanic Immigrants in New York City, 1960–80." In *Hispanics in the U.S. Economy,* ed. George J. Borjas and Marta Tienda. New York: Academic Press.

Serrin, William. 1986. "A Labor Dispute with a Third Side." *New York Times,* January 19: sec. E, p. 5.

Skaggs, Jimmy M. 1986. *Prime Cut: Livestock Raising and Meatpacking in the United States, 1607–1983.* College Station: Texas A&M University Press.

Stull, Donald D., Janet E. Benson, Michael J. Broadway, Arthur L. Campa, Ken C. Erickson, and Mark A. Grey. 1990. *Changing Relations: Newcomers and Established Residents in Garden City, Kansas.* Report no. 172. Lawrence: University of Kansas Institute for Public Policy and Business Research.

Weingarten, Paul. 1988. "It's Not a Stampede, but Beef Is Coming Back." *Chicago Tribune,* January 3: sec. A, p. 5.

6

The Politics of Globalization in Rural Mexico: Campesino Initiatives to Restructure the Agricultural Credit System

David Myhre

Across rural Mexico, global economic and political conditions are inducing many campesinos to attempt to restructure their relations to the Mexican state and to local, national, and international agricultural markets.[1] With rising frequency, campesinos are encountering unfamiliar forms of agricultural finance, production, marketing, and processing that challenge long-standing patterns of their economy and society. Campesinos' efforts to obtain livelihoods from the cultivation of their lands have been disrupted by the quick erosion or disappearance of policies, programs, and agencies that previously supported, protected, and conditioned the development of the agricultural sector. Local markets responsive to local supply and demand conditions, or highly influenced by government price supports and controls, are being disordered and even replaced by liberalized markets of international dimensions

While conducting research in Mexico, I benefited from conversations with the directors and staff of the National Association of Credit Unions of the Social Sector (Associacion Nacional de Uniones de Credito del Sector Social), as well as with the leaders, administrators, and members of several campesino credit unions. I acknowledge financial support for this research from a doctoral fellowship granted by the Inter-American Foundation and from a travel grant awarded by the Center for International Studies at Cornell University.

[1] In this chapter, *campesino* and *campesinado* are used in place of such terms as *peasant* and *peasantry, family farmers,* and *small farmers. Campesino* refers to persons in rural households who cultivate land mainly through household labor or mechanisms of labor exchange. *Campesinado* is employed to refer to campesinos as a collectivity.

that are responsive to price signals from commodity futures markets. In the face of this deep restructuring of the political economy of the Mexican agro-food system, many campesinos are devising new forms of association in order to conserve their traditions and to improve their lives and livelihoods.

Campesinos control access to more than half of Mexico's total land areas.[2] In this vast rural expanse globalization is being mediated through encounters between the state and the campesinado over internationally inspired agricultural policies and programs. And globalization is being felt as well in local markets affected by the liberalization of controls on input prices and the importation of competing commodities. Consequently, since the early 1980s global economic conditions and ideological stances have emerged as key to understanding both campesino movements and agricultural policies in Mexico. Nevertheless, our understanding of how rural politics—including the relationship of the campesinado to the state—is being reconstituted in response to globalization is still limited. Most research to date has concentrated on charting the manifestations of globalization in rural productive and distributive activities (see Barkin and Suárez 1985, Sanderson 1986) or on delineating the nature of state responses to international financial and political institutions regulating the participation of Mexico in the global agricultural economy (see McMichael and Myhre 1991). This focus has been to the detriment of exploring how globalization contextualizes campesino organization and political action. Campesinos, through their organizations, are influencing the form and velocity of globalization in Mexico as they contest with government officials the specifics of new agricultural policies and programs judged to be compatible with international commodity production and marketing arrangements.[3]

The reorganization and shrinking of the official agricultural system have been central to efforts to redirect the Mexican agricultural sector

[2] The available data on average amounts of land held by individual campesino households often are incomplete or contradictory. Nevertheless, there is general agreement that the Mexican campesino household rarely controls more than ten hectares of arable land, and frequently much less. Typically, farming provides a campesino household with only a portion of its income, with the remainder obtained from local wage employment or earnings received by household members during temporary labor migration to other rural areas or to cities in North America. Nonwage income is obtained through barter and exchange or exchange of services and goods or from money sent by relatives who have emigrated permanently.

[3] I concentrate in this essay on regional and national politics rather than macro-level dynamics of global economic restructuring. See the contributions in Long and Long 1992 and Long 1992a and 1992b for the theory and practice underpinning the melding of an actor-oriented perspective with structural analysis.

in the context of an open economy. Agricultural credit has long been a major axis of state-campesino relations (Rello 1987), and access to it has been a recurring motive for campesino mobilization. Through much of the 1970s and 1980s approximately half of all campesinos had access to official credit (Myhre 1991), but since 1989 access has contracted significantly and now fewer than 20 percent of all campesinos obtain official loans. In the aftermath of this abrupt drop in loan availability, access to appropriate forms and adequate amounts of agricultural credit has become a key point of contention between campesino organizations and the Mexican state.

The Origins of Instability and Change in Rural Politics

The origins of the shifting relations in rural Mexico are found in the economic crisis that erupted in full force during the summer of 1982, when the country failed to meet its international debt service obligations. The accompanying economic decline spawned a fiscal crisis that undermined the long-standing model of development predicated on state intervention in markets and economic nationalism. In its aftermath a new generation of elite state officials took power and embarked on establishing a development model characterized by a reduced role for the state and an open economy. Their efforts to curb the regulation of the economy by the state were strongly backed by transnational financial and trade agencies—such as the World Bank and the GATT, and by states, financial institutions, and corporations in the North.[4] To overcome the crisis these officials of neoliberal outlook sharply reduced fiscal expenditures for agriculture, including spending for credit, infrastructure development and maintenance, research, pest control, extension, and, most important, price supports—all elements formerly central to state-campesino relations.[5] Furthermore, they have carried

[4] Mexico joined the GATT in 1985. The proposed North American Free Trade Agreement between Mexico, the United States, and Canada is another example of a transnational trade agency that implies the diminution of national regulation of the economy.

[5] Robles (1992) shows that public investment in the agricultural sector fell approximately 70 percent in real terms between the periods 1980–82 and 1989–91. But the fall has been somewhat offset since 1991 by a significant increase in informal production credit channeled to the most marginal areas by the government's National Solidarity Program, as well as by additional public investment aimed at countering opposition to the November 1991 to February 1992 reforms of the agrarian law first set in the 1917 Constitution. For an initial discussion of these reforms, see Cornelius, 1992. By late 1992 it had become evident that the new codes' effects on agriculture would be slow to be felt (Shwedel 1992).

out a series of macroeconomic adjustments unfavorable to the agricultural sector; in particular, the overvaluation of the peso has facilitated competing imports.

Perhaps nowhere else has the attempt to integrate into global markets been made more forcefully than in Mexico, where policies protecting home markets and supporting their national suppliers have been weakened or discarded and replaced by measures designed to drop producers and laborers into competitive global markets. In this new policy environment, state agencies have fewer resources to spend on the countryside. Consequently, state-campesino relations are being reforged as the state's ability to intervene in or to regulate the campesino economy is weakened. Among campesinos, this situation has triggered a growing perception that the postrevolutionary state—which heretofore had emphasized attention to rural interests at least in its rhetoric, if not in its budgets and actions—is becoming unconcerned with the mounting challenges to campesino livelihoods.

The globally induced shifts in state-campesino relations have induced unprecedented political interactions—namely, negotiated or shared planning of national policies and local programs—between campesinos and the state. Shared planning also responds to the growing belief of campesino leaders that the campesinado will not attain control over the economic structures central to sustainable rural livelihoods as long as campesino organizations are subject to the whims of tutelary state agencies. Throughout the 1980s, growing numbers of campesinos responded to both the antiagriculture policy environment and state tutelage by establishing campesino organizations whose common objective is the implementation of political and economic strategies based on self-managed rural development and self-directed participation in markets (Gordillo 1988b, Fox and Gordillo 1989, Moguel, Botey, and Hernández 1992). These organizations usually reject explicitly the corporatist structures long characteristic of state-campesino relations in Mexico. Instead of meekly submitting to the direction of state officials in matters of rural development, the organizations seek to negotiate for policies and programs they deem could contribute to improving rural living standards and livelihood options.

Since the mid-1980s the concept of globalization—namely, the organization of production relations in the context of competition in global markets—has entered into the everyday discourse of campesino leaders and many campesinos. It is in the context of the growing linkages of local commodity distribution channels to regional and national markets sourcing both domestically and internationally that the new organizations must struggle to find new paths to rural development and sustainable livelihoods and to become identified as actors instead of respondents. It is widely understood, even in remote areas of rural Mexico, that production must adhere increasingly to the discipline of quality standards, and that it must be efficient enough to compete in deregulated markets characterized by falling prices.[6] The self-directed approach taken by the new campesino organizations is premised on the belief that greater control of the productive, transformative, and distributive processes (or circuits) will permit greater efficiencies that in turn translate into lower costs and wider profit margins. One of the characteristics of the globalization of agriculture, however, is that local and regional markets become subject to a wider array of influences that can produce greater volatility in prices and, hence, greater risks. Because the ability of the state to intervene in markets has diminished, campesino organizations face alone a complex market for which their experience has ill prepared them. In short, the complexity implicit in the replacement of domestically controlled markets by globally sensitive markets raises obstacles to effective participation by campesinos in the restructured agricultural economy.

The Crisis of Rural Mexico

Most observers of the Mexican agro-food system concur that it entered deeply into crisis in the early 1970s and that little if any progress in resolving the crisis has been achieved. The primary characteristics of this persistent predicament include declining per capita production of basic food crops, mounting rural-urban migration, increasing levels of child malnutrition in the countryside, rising indices of landlessness and rural unemployment, deteriorating agro-ecological endowments, and falling levels of investment and farm capitalization (see Barkin and Suárez 1985, Sanderson 1986, Barkin 1990, Robles 1992).

The success of the Mexican government in the early 1990s in stabilizing its internal and external debt, in attracting repatriated capital and

[6] I found a striking example of this trend during a visit to an indigenous community. A community leader, about 35 years old, explained to me that in the future the corn he and his fellow campesinos grow will have to adhere to standards of "total quality" and be efficiently produced if he is not to be knocked out of the "global market." He said, "Globalization is on its way. God willing, we will be able to improve our productivity and survive." His view may be overly optimistic, since yields on the better cornfields of Mexico now are three times higher than those in his community, and yields in Iowa are five times higher.

new foreign investment, and in stimulating renewed economic growth in the range of 3 to 4 percent annually has been interpreted by some—especially neoliberal state officials with little direct experience in the agricultural sector—as a signal that the long-awaited recovery of the rural economy would begin. But observers, as well as the members and leaders of the independent campesino organizations that have gained strength in Mexico since the early eighties, have argued that the campesinado is not sharing in the alleged economic revitalization (Robles and Moguel 1990, Carlsen and Robles 1991, Hernández 1991, Robles 1992, Shwedel 1992). Campesino organizations frequently point out that the campesino economy is being decapitalized as farm households find it impossible to finance the use and maintenance of yield-enhancing technologies, ranging from agrochemicals to tractors. In this context, households are forced to reduce the area and density of planting and revert to more labor-intensive production techniques that usually result in lower yields, hence lower incomes. These downward-spiraling pressures are exacerbated by falling real prices associated with the withdrawal or reduction in state subsidies and with the increasing availability of cheaper agricultural imports. Furthermore, because the state is ceasing to support the campesino economy and campesinos have less capital, investment in the infrastructure needed to compete over the long run with the subsidized agricultural sectors of the advanced industrialized nations is sorely missing.

The Revitalization of Mexican Campesino Organizations

The current configuration of state-campesino relations can be traced to the first half of the 1980s, when a number of newly formed campesino organizations began to articulate proposals designed to lessen state corporatist mechanisms of control—such as official campesino unions, closed marketing arrangements, and supervised credit—over campesino farmers.[7] These regional and national organizations differentiated themselves from other campesino organizations not only by mobilizing protest actions demanding the recognition and fulfillment of agrarian rights fought for, but not fully won, by campesinos during the Mexican Revolution, but also by articulating proposals for new forms of campesino economic organization and of state involvement in the rural

[7] The history of the formation of these new organizations is excellently reported by Neil Harvey (1990).

economy. The new rural popular organizations, encountering a state framework in the throes of change and reform, were able to utilize their newfound capacities to design and propose alternative policies to attract members tiring of the corporatist approaches to rural politics. The context of the shift from a command to an open economy was propitious for mobilization to demand an end to tutelary practices by state agencies and to corporatist politics.

The strategy of engaging in a "politics of proposal" instead of a "politics of protest" facilitated participation by the organizations in decision making, formerly the sole province of state officials (Gordillo 1988a, 1988b; Fox and Gordillo 1989; Hernández 1990a). Indeed, several of the organizations' proposals for agricultural policy reforms and the reorganization of state agricultural agencies had been partially implemented by early 1993, especially in the area of the reorganization of state marketing of agricultural inputs and outputs. In the context of the economic restructuring gripping Mexico, however, these new rural popular organizations—like many of their urban-based counterparts—today face the dilemma of how to continue to develop innovative answers to the economic crisis that is battering their members while maintaining their organizational principles and cohesiveness (Hernández 1990b).

UNORCA: Building the New Campesino Organization

In the early 1980s a series of informal meetings promoting dialogue among the memberships of local and regional campesino organizations was organized by leaders of campesino organizations concentrated in northwestern Mexico. The talks centered on the necessity of gaining greater autonomy in the production process and retaining greater control over the profits generated in the marketing chain.[8] On March 31, 1985, the meetings culminated in the formation of the National Union of Autonomous Regional Campesino Organizations (UNORCA) by twenty-four campesino organizations from fourteen states. By early 1990 UNORCA was among the largest of the national campesino organizations, with its member rolls comprising over sixty regional campesino organizations and unions.[9]

[8] The high degree of state control over campesino production through the regulation of credit, extension, and marketing controls has been widely documented (e.g., Gordillo 1988b) and will not be examined here.

[9] The Coordinadora Nacional Plan de Ayala and the Central Independiente de Obreros Agricolas y Campesinos are foremost among the other major independent national agrarian popular movements.

Since its founding, UNORCA has been very influential in shaping the emphasis on self-managed regional economic development held by major streams of the campesino movement.[10] Its efforts to build horizontal links between regional and national campesino organizations, including those allied with the corporatist centralized campesino confederations connected to the ruling Party of the Institutionalized Revolution (PRI), and the Mexican state have defined new pluralistic practices in agrarian politics. Unlike centralized campesino organizations (both independent and corporatist), where leaders typically isolated from both the rank and file and from regional and local leaders develop political strategies without input from the base, UNORCA has been able to draw from its member regional campesino organizations a corps of capable leaders with grassroots contacts. These regional leaders are better able to construct national proposals that reflect diverse regional realities than are leaders of centralized organizations. In so doing, they are more likely to elicit broad support because they more directly reflect the demands of local leaders and rank-and-file members.[11] Through a long series of working sessions and national plenary meetings in which proposals and political strategies are developed, UNORCA's leaders usually have been able to work effectively together. After formulating a policy position, they then forcefully articulate it to the public and mobilize support among the memberships of their respective regional organizations.

By virtue of focusing on rural economic growth rather than the politically charged issue of land redistribution, UNORCA's proposals do not spark an immediate defensive reaction from the state, which, after all, has development on its agenda. Furthermore, UNORCA's proposals commonly call for shifting control over economic institutions from the state to the campesinado and thus appeal to those state officials who perceive such changes to the state's role in the countryside as leading to a reduction in fiscal outlays.

The support of state officials for the retrenchment of the state in the rural economy deserves attention, for it reflects how factors beyond the full control of state leaders have opened a contradictory space in which groups such as UNORCA can operate. The Mexican state has been pressured by international development finance institutions to reduce

[10] Indeed, the dominant ruling party-affiliated campesino organization, the National Campesino Confederation, is now led by a former UNORCA strategist who is trying to implement programs supporting greater economic self-organization by regional and local campesino groups. For additional detail on this development see Hernández 1992b.

[11] I am grateful to Luis Hernández for discussions on this point (also see Hernández 1992a).

its involvement in the countryside. Given its limited fiscal resources, the state must turn increasingly to campesino organizations for assistance in meeting another demand of the international financial institutions, namely, the reorganization of the agricultural economy. There is potential for conflict when the pressure to withdraw results in a lack of fiscal commitment to assist the campesino organizations in the initial period of their takeover of former state functions (e.g., the operation of parastatals), or to provide some subsidies to producers facing harsh markets. In other words, although the pressures from within and without on the state to withdraw from the rural economy complement UNORCA's overall goals, they also constrain UNORCA's potential opportunity to extract the resources its member organizations need to handle the normal costs of transition, not to mention the suddenly heightened competitive pressures associated with the opening agricultural economy.

Assessing the Experience of UNORCA

The nature of the coalition of organizations composing UNORCA and the strategies they use to represent their interests signal important changes in state-campesino relations and in the Mexican campesino movement. UNORCA has opted for a strategy of consideration of the state's overtures, although it can and does oppose the state on particular policies. By negotiating with the state, UNORCA apparently is responding to its organizational imperative to influence agricultural policies and to gain control over state fiscal resources. In so doing, however, UNORCA helps the state in its attempts to legitimate itself vis-à-vis the campesinado and in its efforts to justify its current neoliberal prescriptions for the agricultural economy. Consequently, UNORCA is in a dangerous position: on the one hand, it may be overdependent on support from a state fundamentally uncommitted to the campesinado, and on the other hand, it risks the state's advertising UNORCA's connection to it. Such advertising could undermine its oppositional stance, which is a major organizing tool in attracting support from the campesinos who feel the state is not their ally.

UNORCA, by willingly negotiating and forging compromises with the state, risks having its actions viewed as tacit support for the plethora of reforms being promoted by the state—such as cutbacks in rural investment and credit—which are profoundly unfavorable to campesinos. These reforms disguise the state's partial withdrawal from the economy. The very liberalization policies that are being legitimated, however, are effectively transforming the terrain upon which UNORCA struggles: the privatization of parastatal agencies and the liberalization of markets arguably decrease the state's ability to act according to UNORCA's interests. In addition, UNORCA's desire to foment autonomous control by its own membership receives support from the state, but this support is predicated on the state's ceding control of key economic institutions to powerful private forces that may oppose UNORCA's goals. Furthermore, UNORCA's ties to other campesino organizations preferring to protest, rather than to negotiate, agricultural sector policies—especially those related to land access—are weakening as years of the state consolidates its neoliberal project. As that occurs, officials could be emboldened to mobilize state resources to repress the old-line campesino movement built up around demands for land instead of market reform.

But UNORCA has within it the seeds to overcome these challenges to its autonomy and to its ability to mobilize in defense of the campesinado. Its decentralized structure precludes easy incorporation into the state's ranks, for cautioning and opposing voices can be heard, including those tied to oppositional political parties and movements (Hernández 1992a). Furthermore, if its member organizations achieve economic success and lower their reliance on state fiscal contributions, then UNORCA may gain greater independence from the state.

The Development of the Campesino Credit Union Movement

Since about 1978 various independent campesino organizations have established firms to distribute farm inputs and to market farm production as part of their efforts to obtain greater control over the productive process and to retain more of the economic surpluses generated by their members' labor. The self-managed firms recognized early that reliable access to timely and adequate credit was necessary if members were to be successful in supplying quality commodities to their new enterprises. Official agricultural credit institutions were unattractive loan sources because of their consistent use of lending mechanisms and calendars that wrested production and marketing decisions from borrowers—anathema to the autonomy credo of the new organizations. Meanwhile, the commercial banking sector exhibited little interest in lending to all but the most promising of the emergent independent campesino enterprises. Given these circumstances, proposals to establish alternative financial institutions attentive to the needs of the social sector began to circulate among leaders of the new types of campesino organizations—especially within UNORCA.

The Context for UNORCA's Efforts to Establish Credit Unions

Mexico's legacy of economic development based on the use of pricing mechanisms to extract capital disproportionately from the countryside for urban and industrial investment, combined with the capital-draining impact of the debt crisis, resulted in a shortage of investment in and credit for the agricultural sector. Without long-term capital improvement and short-term crop production loans there were few prospects for improving the contribution of the social sector to the supply of basic grains or for alleviating the deteriorating socioeconomic conditions afflicting the third of Mexico's population who live in the countryside.

Clearly, agricultural credit is a principal axis of state-campesino relations. It also is a key source of liquidity in the crisis-ridden rural economy. As such, agricultural credit plays an important role in dampening crisis-induced rural unrest, as well as opposition to the government and the PRI. Concern for the political fallout, as much as for negative effects on productivity, likely has prevented the Mexican state from completely shutting down the pipeline of cash to the rural economy, which the credit system became in the late 1970s and early 1980s (see Pessah 1987, Myhre 1991). Policy makers have realized that the state must continue to promote agricultural credit as one of its primary policy instruments for addressing the needs of rural people. The fiscal crisis of the Mexican state during the 1980s, however, resulted in a decline in overall lending to campesino producers.[12]

It was against this backdrop that policy makers began to recognize the need for, and UNORCA began to make proposals about, the restructuring of the rural finance system. Carlos Salinas de Gortari, during his 1988 presidential campaign, and later upon taking office, made reform of the budget-draining agricultural credit system a central element of his administration's plans for reactivating the moribund agricultural economy. He implemented many policy changes that generally reduced state outlays for investments in rural infrastructure, severely cut subsidies on most commodities, and curtailed lending. The latter

[12] Interestingly, during the mid-1980s, as real total lending tended to decline—in accordance with the necessity of alleviating the fiscal crisis—total area financed actually increased—in accordance with the need to provide both productive and political stimulants (see Myhre 1991).

action reflected new plans for the rural financial system, particularly Banrural (the official agricultural credit bank), which was to be placed on a businesslike footing and cease lending to farmers with limited or no productive potential or with a history of not repaying their loans. These actions resulted in a 70 to 80 percent decline in Banrural lending when compared with its peak of activities in the early 1980s.[13]

To mitigate this sudden and sharp contraction in loan funds available to campesinos, there were expressions of official support (including statements by President Salinas) for the formation of credit unions to service campesinos deemed to be cultivating land with moderate to high productive potential. In addition, the formation of credit unions responded to the state's initiative to reduce its size, since the cost of a credit union's operations is mainly absorbed by its members. Thus beginning in early 1989 state agricultural policy makers responded cautiously to demands by UNORCA to help facilitate the establishment of credit unions by providing some support for technical assistance and training. The restructuring of the rural credit system has emerged as one of the main areas where the state is redefining its role in the rural economy and its relationship to the campesinado.

UNORCA's Credit Union Drive

In response to the mid-1980s collapse of official agricultural credit for campesino farming,[14] UNORCA's member organizations set as major organizational tasks lobbying for agricultural credit policies more favorable to campesinos and establishment of credit unions. Much lobbying focused on changing banking-sector regulations that impede the establishment of credit unions by regional campesino groups. Regulations concerning credit unions originally were designed to facilitate the formation of large industrial credit unions by members able to make sizeable contributions to reserve and operating capital. UNORCA's organizational commitment was required in order to assist regional campesino organizations seeking to form credit unions with negotiating the regulatory framework.

Since 1986, and especially since late 1988, UNORCA has used

[13] The curtailment is not quite so drastic, however. Most of the campesinos who were excluded by Banrural are being serviced by the neocorporatist National Solidarity Program, which is providing them with "soft" loans made "on the campesino's word of honor." These loans typically are about one-quarter of the comparable amount per hectare lent by Banrural.

[14] Official credit disbursed fell 43 percent from 1985 to 1987 (Banamex 1989), and the downward trend continued into 1992, as documented by in Correa 1989, and Correa and Robles 1990, and Robles 1992.

contributions from official and nongovernmental donors, and funds raised by regional campesino organizations interested in forming credit unions, to contract staff to help negotiate the regulatory maze, form plans for credit union operations, and to help set up the credit unions (see Cruz Hernández and Zuvire Lucas 1991). By late 1989 ten credit unions had been authorized, of which eight were able to begin operations, by late 1990 twelve credit unions were authorized, of which ten were functioning.

As the Salinas administration got under way in 1989, the fledgling campesino credit unions faced difficult challenges, especially that of remaining capitalized in an agricultural economy characterized by price and climatic variability. UNORCA confronted this destabilizing situation by entering into direct negotiations with state officials in an attempt to obtain technical and financial assistance for the credit unions. In February 1990 UNORCA's Finance Committee and member credit unions successfully negotiated an agreement with seven state institutions for just such assistance to be channeled to ten credit unions. As part of the agreement, a credit line amounting to approximately US$3.5 million was negotiated with the official agricultural credit institutions, especially Banrural, to augment the credit lines obtained in individual negotiations by the credit unions. This agreement is a concrete example of UNORCA's strategy of negotiation and participation in the process commonly known as *concertación* (shared planning), but it also presents—as noted earlier—the potential for the establishment of dependency on the state. Concern about dependency on the state began to fade, however, as the credit unions faced an unheralded branch of the state, namely, the financial system regulators.

The State and the Campesino Credit Unions

Even as the Salinas administration pushed the institutions of the official agricultural credit system to open up to campesino credit unions, there was growing evidence that UNORCA's ability to mobilize campesinos to politically negotiate access to credit did not then lead to an easing of the restrictions on that access. Political mobilization had little effect on the application of First World financial accounting criteria and banking procedures by the outwardly apolitical banking regulatory agencies responsible for vigilance over the operations of the comparatively rustic credit unions. The leadership and staff of these agencies were concerned mainly with creating a globally competitive banking system; they had little interest in agrarian politics, and less regard for campesino activists as bankers. The technical complexity of negotiations with the banks and the financial regulatory agencies, as well as the needs of the credit unions for specialized technical assistance, evidenced the utility of establishing an organization of campesino credit unions that would remain largely aloof from campesino politics. By establishing a separate entity, the credit unions could better represent themselves and provide mutual operational support. Therefore, in October 1990 the UNORCA-supported credit unions issued a call to other independent campesino credit unions to meet and discuss the formation of a network focused strictly on issues of financial policies and regulations. At the meeting in November 1990 was born the National Association of Credit Unions of the Social Sector (ANUCSS).[15] ANUCSS came from the alliance of the fourteen existing UNORCA unions with another nine campesino credit unions pertaining to other coalitions or independent organizations of the social sector.

In Mexico, where the organizations that make up the new rural social movements have had relative success at influencing leading policy makers, the organizations' ability to maneuver in the bureaucratic environment is emerging as a key determinant of their success or failure; it is precisely in the offices of frontline state agency officials that the relationship between the Mexican state and the campesinado, negotiated by politicians and campesino leaders, actually is being restructured. I believe that the organizations' bureaucratic savvy is even more important when the nexus of state-campesino interactions is located in regulatory bodies, rather than in the offices of state agencies responsible for implementing programs in the countryside. There is less congruency of interests between regulators and the regulated, as the former typically adopt an adversarial position to the latter, which is only exacerbated by their low expectations for the capacities of campesinos and credit union staffers new to the maze of the financial system. Whereas implementing agencies with responsibilities for programs and services are filled with bureaucrats who must at least nominally regard the campesinos as constituents or clients, regulatory bodies are not.

The Politics of Implementation

Although many leading officials of Salinas's economic and agricultural cabinets have offered rhetorical and material backing for the for-

[15] In Mexico "social sector" refers to rural social collectivities known as *ejidos* and to indigenous communities. In both cases, the government grants campesinos long-term usufruct to

mation of campesino credit unions, they apparently have limited capacity to ensure that the regulatory agencies establish a balance between guarding the health of the urban-oriented financial system and support for the evolution of credit unions in the harsh rural environment. Understanding how and why some elements of the Mexican state adopt obstructionist positions vis-à-vis the campesino credit union movement is critical both to explaining the latter's fitful progress and to recognizing the importance of the formation of specialized or sectoral organizations such as ANUCSS within the campesino movement. These specialized organizations represent a new level of campesino organizational capacity, able to work with, and, occasionally, to confront, the regulatory agencies. In so doing, they are engaging in a "politics of implementation" whereby campesino-state discussions focus on modifying the specific aspects of agricultural policies and programs and the procedures used to implement them. This contrasts sharply with the longstanding pattern of the "politics of mobilization" whereby campesino organizations draw on agrarian principles rooted in the Mexican Revolution when issuing general demands for state support for the rural sector.

In the past, agrarian organizations were established to battle for land reform and other items associated with the historical agenda of the revolution. They also have been constructed around new "public goods" issues, such as rainforest preservation. In these circumstances, the organizations can adopt and argue from a moral perspective and work out compromises with policy makers and officials operating in a context where the rules of the game are fluid and open to negotiation. As the agrarian organizations mobilize around investment and banking issues, however, they confront officials whose outlook is shaped by the need to establish stability in the financial system and who are less responsive to arguments framed in terms of the historical commitment of the state to the campesinado. Thus as the politics of mobilization become less appropriate in the increasingly rulebound atmosphere of neoliberal reformism, then the ability to carry out a politics of implementation will become increasingly relevant to the efficacy and survival of the economic projects of the agrarian organizations.

First as the UNORCA credit union project, then later as ANUCSS,

land under terms of the Mexican constitution. A small percentage of the members of the credit unions grouped in ANUCSS are private property owners—known as *minifundistas*—who possess amounts of land comparable to those controlled by members of the social sector.

In early 1992 ANUCSS changed its name to the Mexican Association of Credit Unions of the Social Sector (AMUCSS), but in this chapter the former usage is employed.

the representatives of the campesino credit union movement frequently have attempted to convince the financial regulatory agencies to recognize that the circumstances of financial activity among campesinos—namely, movement of small amounts of money, financially unsophisticated members, and geographic dispersion—compound the complexity of administering a credit union to the letter of the financial regulations. In the face of this unique context, they have argued for the relaxation of rules designed basically for credit unions with financially sophisticated members (e.g., small to large industrialists) and with staff experienced in the nuances of compliance to banking statutes and reporting requirements. They contend that campesino credit unions are in their infancy, and that they must crawl before they will walk. Like a parent who can stifle a child's development by restricting its freedom to commit mistakes and learn from its errors, the regulatory agencies are viewed as applying standards inappropriate to the financial capacities and needs in the countryside. The strict regulations are counterproductive, since adhering to them consumes time that staff and leaders of the credit unions could devote to both supporting members' productive endeavors and to strengthening members' understanding of and commitment to the self-managed financial institution that they have joined.

Confronting the Regulatory Maze

Perhaps the foremost example of how bureaucracies and bureaucrats can affect the chances for successfully achieving the goals of the credit union movement is found in the interactions between those promoting or directing credit unions and the National Banking Commission (CNB), which regulates the Mexican financial sector. The CNB is a line agency managed by career officials steeped in financial accounting principles and has remained rather insulated from control by political appointees responsive to presidential initiatives. The CNB supervises the formation and operation of credit unions, and hence is the site where campesino financial activists encounter the major challenges to their proposals for a new model of rural credit. It has long defined its primary mission as the regulation of the commercial banking sector and has devoted most of its material and human resources to that end.

Unlike other key officials and agencies who are at least conversant with the changes in the agrarian structure and the agricultural economy in Mexico, the CNB and its functionaries long have been isolated from the problems of the countryside. As a result, CNB regulators lack knowledge of the range of economic, social, political, and agro-ecological factors that differentiate the operations of campesino credit unions from those of larger industrial credit unions or commercial banks. This blindness has led them to be unbending in their attitude that the campesino credit unions—most of which are lending US$200,000 to $500,000 per year—should be treated the same as the huge commercial banks with decades of experience at, and the resources for, negotiating the CNB's regulatory maze. When campesino credit union directors and staff attempt to communicate with the officials of the CNB, they discover they must devote much attention and energy to framing experiences and constraints in a manner that the officials can integrate. In other words they must learn the language of the CNB officials.

The campesino directors and the administrative staff of an aspiring credit union must comply with a series of stringent rules and complex procedures promulgated by the CNB to regulate the formation of financial institutions generally, and credit unions particularly. At least fourteen major steps must be completed sequentially before the CNB extends authorization for operation to a credit union. Of course, it is unreasonable to expect that a campesino organization should be able to just walk into the offices of the CNB and start up a credit union. But the CNB's demands for feasibility studies, detailed lending plans, and evaluations of the potential directors and staff—among other requirements—result in great expense, much red tape, and lengthy delays for the campesino organizations backing the credit union project. Although ANUCSS has developed some expertise in guiding credit union projects through the maze of CNB regulation, it is not uncommon for feasibility studies and lending plans to be rejected once or twice before receiving final approval, thus often nearly two years can elapse before approval to operate is gained. Meanwhile, the sponsoring campesino organizations face the difficult task of maintaining enthusiasm among their memberships for a project that seemingly is not imminent, as well as incurring the costs associated with processing applications, promoting the potential credit union, and paying staff salaries long before they can be recovered from revenues obtained in lending operations. Through early 1993 the CNB has made no conciliations to the credit unions that would facilitate the realization of the commitments made by the president and his cabinet members. Its appoach to the campesino credit unions does not foment their self-development—early credit union gains have been made in spite of, not because of, the CNB.

Further evidence of the pivotal role a regulatory bureaucracy such as the CNB can play in blocking the efforts of campesino organizations to become economic actors in regulated markets is presented by a recent series of grave threats to the growth, survival, and consolidation of the campesino credit union movement. In March 1991, without previous consultation, the CNB quadrupled the amount of funds (from about US$50,000 to $200,000) that the membership of each credit union must deposit in a fixed social capital account (roughly equivalent to a bank's reserves account). The credit unions were given until August 31, 1991, to comply with the new capital requirement, which was followed by another hike to US$400,000, effective at the conclusion of 1992. Both the first and the subsequent increases in the fixed social capital account challenged the financial resource mobilization capabilities of the campesino credit unions, whose scope of operations are not comparable to the larger and better-capitalized credit unions found in the private sector and for whom the new rules were less onerous. The member credit unions of ANUCCS accepted the logic of increasing the capitalization of each credit union but opposed the immediate application of a universal criterion, instead arguing for gradual increases in capital over a period of several years and in accordance with the size each credit union's operations. The abruptness, magnitude, and short amount of time the regulatory agency assigned for the realization of each increase in capital funds contrasted clearly with the supposed policy of *concertación* on decisions affecting campesino livelihoods and institutions that agricultural policy makers had extolled.

There also are aspects of the restructuring of the Mexican economy and its internationalization that impinge on the prospects for ANUCSS and its member unions to negotiate the regulatory maze. One of the major initiatives of the Salinas administration, the reprivatization in 1992 and 1993 of the commercial banks nationalized in 1982, may have had an adverse impact on the ability of the campesino credit unions to gain the attention and understanding of the CNB. In early 1992, the CNB began devoting most of its energies to designing and implementing the reprivatization program, while in 1993 it focused on monitoring the newly privatized banks. Thus its tendency to not address the particular problems of the campesino credit unions was reinforced by its need to devote most of its human resources to this major reshaping of the financial sector. A CNB official also explained that the capital requirement for credit unions was raised as part of an attempt to improve the capitalization of the financial sector in order to strengthen it before permitting competition from international banks. Though such a cautious approach indeed may be sensible protection for the large commercial credit unions and banks, it certainly is less relevant for the campesino credit unions, which serve mainly producers in a portion of the agricultural sector characterized by unattractive low profits. Given the pace and magnitude of the reorganization of the major elements of the Mexican financial system, it is not difficult to understand how the comparatively low-profit campesino credit unions can get lost in the regulatory shuffle.

Overcoming the Legacy of Inefficiency

If relations with the CNB have been marked by mutual incomprehension, then relations with Banrural (the major official agricultural credit institution for the campesino sector) can at best be characterized as difficult, and at worst as mutually antagonistic. Although the size and scope of Banrural have been greatly reduced, it still influences the credit unions because they are highly dependent on it as a source for funds to re-lend to their members.[16]

One of the major conflicts in the relations of the credit unions with Banrural revolves around the terms of remuneration to the credit unions for the intermediation services they provide in place of Banrural. When the credit unions borrow a large sum for subsequent relending to their members, it is the official Agricultural Trust Fund of the Bank of Mexico (FIRA), and not Banrural, which is usually the ultimate source of the borrowed money. Through its rediscounting fund, FIRA provides the bank originating the loan (i.e., Banrural) with the actual loan funds at a low interest rate, as well as a loan guarantee covering 80–90 percentage of the amount borrowed. Hence, Banrural is engaged in low-risk lending and earns profit on the margin between the interest rate on the discounted funds and the amount lent. FIRA also provides Banrural with six "points" (i.e., 6 percent of the amount lent) for its intermediation activities. Banrural, as a matter of policy, refuses to share these points with the credit union, despite the fact that the latter is incurring the majority of the intermediation costs associated with servicing individual producers. The credit unions and ANUCSS attempted for three years to obtain a portion of the intermediation points from Banrural.[17]

[16] Campesino credit unions primarily depend on loans from official and private banks for the funds they lend to members. Banking regulations permit credit unions to borrow up to thirty times the total amount of the capital contributions made by their memberships. The credit unions would prefer, however, to lend funds they have mobilized in their own communities. Unfortunately, regulations severely restrict the credit unions' ability to extend membership to agricultural laborers and other rural residents, thus blocking access to a large number of potential members who seek security for their savings rather than access to loans.

[17] The commercial banks also do not share the intermediation points, but because few credit unions are operating with them, their refusal has not become an issue for ANUCSS. Those credit unions that borrow from the commercial banks, however, report that faster and more comprehensive loan servicing and access to other banking services—including checking accounts and project planning—tend to make them less anxious to press on this point.

In late 1992, finally bending to lobbying by ANUCSS, Banrural agreed to share up to four of the six points with the credit unions if they fully realized a series of intermediation tasks with *each* individual member-borrower. By mid-1993, however, only one member credit union of ANUCSS had received reimbursement under this scheme, and then for just one point.

With reasoning indicative of its lukewarm support for the campesino credit unions, Banrural justified retaining the intermediation points on the grounds that, for it, the credit unions were borrowers like any other borrower—albeit large—and just, as they would not share points with an individual farmer or group of farmers, they would not share them with the credit unions. Under the terms of the financial laws, however, the credit unions are distinguished from individual borrowers by their status as "auxiliary credit organizations" able to carry out financial intermediation activities in partnership with banks in the financial system. Despite this, Banrural's resistance continues, as evidenced by how the new point-sharing scheme downplays the role of the credit union as an independent financial institution and instead tends to reduce it to the status of a conduit between individual borrowers and Banrural. This situation is a product of the requirement that intermediation activities eligible for point-sharing be evaluated for each member-borrower, rather than through evaluation of the overall intermediation success of the credit union (e.g., amount lent, timeliness of loan disbursements, and repayment rates). Banrural simply has ignored categories set up by the Mexican financial system. There are indications that the profound liquidity problems of Banrural, given its fiscal shrinking, in tandem with cutbacks in state support for agriculture, force it to seek income from all sources. In short, the fiscal crisis of Banrural pushes it to resist cooperating fully with the campesino credit unions. Banrural's interests are best served by putting the credit unions in a position where they are barely surviving. As long as the credit unions are able to survive, Banrural will be able to move large sums of funds for minimal administrative costs (especially since the repayment records of the credit unions are quite good) and at low risk, thus earning profits that can support its less successful loans and underwrite its large staff.

Conclusions

Two major points emerge from this analysis of the campesino credit union movement. First, Mexico illustrates the key role played by state agents in regulating the advance of globalization in rural economy and society. State officials, guided by their own neoliberal economic principles, downplayed domestic political criteria in favor of the exigencies associated with global economic competitiveness. Their efforts were lent legitimacy by transnational finance and trade institutions, which pressured for an accelerated adoption of international standards and prices for agricultural commodities. The resulting move to accept global standards has outpaced the development by the new campesino organizations of their capacities to negotiate changing market circumstances without the state as a supportive partner. The groundwork for some of this rapid change had been laid by years of gradual changes behind protective barriers in the nature of agricultural production and marketing in the campesino sector.

The emphasis on rapidly matching global standards for competitiveness virtually has eliminated time for steering carefully and experimentally to a new course for rural development. Before the opening of the Mexican economy accelerated in 1989, the innovative campesino organizations were already embarked on a course of improving economic efficiency and competitiveness. But they also were guided by a desire to preserve and improve the social conditions of rural life as they made the attempt. Their efforts to respond humanely to the challenge of the restructuring of the Mexican agro-food system have unraveled due in large part to the abandonment by the state of its historical partnership with the campesinado. Raised by nearly seventy years under the watchful eyes of a tutelary agricultural bureaucracy, the campesinado has not been able to achieve independence in a short time with its limited arsenal of material and human resources. The state has contributed greatly to the rapid erosion of its historical pact with the campesinado and has failed to lay firm foundations for a new bond.

Certainly, without the Salinas government's ideological commitment to neoliberal policies the campesino movement still would have encountered pressures associated with the globalization of agriculture. But the adopted neoliberal ideology exacerbated the impact of globalization because it resulted in the restriction of state resources to facilitate the incorporation of campesino organizations into the opening agricultural economy. In short, the prospects for Mexico's campesino movement have been tempered because of the state's decision to act abruptly and thus foment a deepening of the rural crisis, precisely at the moment when the movement for self-managed economic development was just beginning to find its feet. A more orderly entrance into the global economy and a more methodical constriction of state activities could have dampened the pace of change without altering its course. It also would have permitted the gradual building up of the new campesino movement's understanding of markets. The present state-propelled approach, however, has resulted in a premature crisis for the campesino movement, as exemplified in the crisis of the credit unions detailed here. The abrupt pace of change has precluded the incremental capitalization of campesino economic organizations and the gradual building up of trained and competent staff. It is particularly evident that the abilities of campesino movements to "propose" to the state have been severely eroded by the reduction of its presence in the agricultural sector.

A second major point is the importance for campesino organizations of establishing the capacity to carry out the politics of implementation when entering into specialized markets still subject to regulation by the state. In the current context of state restructuring campesino organizations must think about strategies for pressuring regulators to facilitate the distribution of resources and other support won from politicians and policy makers during negotiations about policies and programs. It is clear that regulatory agency bureaucrats, as opposed to field or programmatic agency officials, are the state actors who must translate the "politics of proposal" into actions. Organizations that do not learn to frame their proposals and plans in the language and criteria of regulators are likely to be frustrated in their efforts to innovate new economic activities because regulatory officials can easily convert their own ignorance of agriculture into a brake on efforts to experiment.

The above two points also indicate important developments in the context in which the campesino movement is evolving. Globalization challenges campesino leaders to amplify their understanding of how Mexico's agricultural economy is operating in an international context. They must become experts in the politics of GATT, bilateral trade negotiations, international commodities futures markets, and so forth. Many are already gaining expertise on these issues, but they are being severely pressed to absorb and analyze large quantities of useful information. Concentration on learning diverts their limited energies from their organizing tasks. Technical and administrative staffers face similar challenges in learning the ins and outs of regulations and standard operating procedures. One observer of rural Mexico has warned of a tendency to develop a class of white-collar leaders and technicians, removed from their membership base because of the exigencies of running campesino economic organizations in an environment regulated by concerns for economic efficiency that are far from the social welfare criteria upon which their organizations were founded (Bartra 1991). I fear that campesino organizations, during battles over the policies of implementation, are increasingly being pushed to accept the standards and rules by which private firms carry out their activities. But it is an open question whether these standards are the best way for organizations to achieve an outcome that balances economic and social criteria. As long as the campesino economic organizations are weak, they are vulnerable to such external assaults on their social agendas. The coming challenge is to match a vision of desired rural social conditions with the possibilities offered by the market context in which the campesino economy now must function.

The experiences of campesino credit unions in Mexico raise a cautionary flag to those who see the new forms of campesino economic organization as the best future hope for rural Mexicans. Though it is true that during the mid-1980s they had initial success confronting a gradually globalizing economy and a state slowly reshaping itself, it remains an open question whether they can adapt to the exigencies of the current conjuncture marked by the ongoing paring of the state, the further opening of the economy under NAFTA, limited access to state fiscal resources and political sympathy, and pressures to become more technocratic in their organizational dynamics and criteria. A positive resolution to these problems will depend on whether the leaders and staff of campesino organizations can gain a new range of technical skills that will enable them to better analyze how to do more with less and how to help protect campesino social values as they and their fellow campesinos negotiate new sets of regional, national, and global political and economic relationships.

References

Banamex. 1989. "Consideraciones sobre el sector agropecuario." *Examen de la Situación Económica de Mexico* 65(767): 439–44.

Barkin, David. 1990. *Distorted Development: Mexico in the World Economy.* Boulder, Colo.: Westview Press.

Barkin, David, and Blanca Suárez. 1985. *El fin de la autosuficiencia alimentaria.* Mexico City: Editorial Océano.

Bartra, Armando. 1991. "Pros, contras y asegunes de la apropriación del proceso productivo: organizaciones rurales de productores." *El Cotidiano* no. 39: 46–52.

Carlsen, Laura, and Rosario Robles. 1991. "El Tratado del Libra Comercio y la agricultura." *El Cotidiano* no. 40: 3–10.

Cornelius, Wayne A. 1992. "The Politics and Economics of Reforming the Ejido Sector in Mexico: An Overview and Research Agenda." *LASA Forum* (Bulletin of the Latin American Studies Association) 23(3): 3–10.

Correa, Guillermo. 1989. "Falto de inversión, de crédito, de maquinaria: el campo no produce para que el país coma." *Proceso* no. 668: 14–17.
Correa, Guillermo, and Manuel Robles. 1990. "Sin recursos, sin servicios, sin créditos, hasta los ejidatarios abandonan al ejido." *Proceso* no. 711: 16–21.
Cruz Hernández, Isabel, and Martin Zuvire Lucas. 1991. "Uniones de crédito agropecuarias: una red que viene de lejos." *Cuadernos desarrollo de base* 2: 185–220.
Fox, Jonathan, and Gustavo Gordillo. 1989. "Between State and Market: The Campesinos' Quest for Autonomy." In *Mexico's Alternative Political Futures,* ed. Wayne A. Cornelius, Judith Genteman, and Peter H. Smith, pp. 131–72. San Diego: Center for U.S.-Mexico Studies, University of California.
Gordillo, Gustavo. 1988a. *Campesinos al asalto del cielo.* Mexico City: Siglo Veintiuno Editores.
——. 1988b. *Estado, mercados y movimiento campesino.* Mexico City: Plaza y Valdés.
Harvey, Neil. 1990. *The New Agrairan Movement in Mexico: 1979–1990.* Research Paper no. 23. London: University of London, Institute of Latin American Studies.
Hernández, Luis. 1990a. "Las convulsiones rurales." *El Cotidiano* 34: 13–21.
——. 1990b. "Las coordinatoras de masas: Diez años de trincheras." *El Cotidiano* no. 36: 34–35, 42–46.
——. 1991. "Respuestas campesinas en la época del neolliberalismo." *El Cotidiano* no. 39: 53–58.
——. 1992a. "La UNORCA: Doce tesis sobre el nuevo liderazgo campesino en México." In *Autonomía y Nuevos Movimientos Sociales en el Desarrollo Rural,* ed. Julio Moguel, Carlota Botey, and Luis Hernández, pp. 55–77. Mexico City: Siglo Veintiuno Editores.
——. 1992b. "Las telaranas de la nueva organicidad." *El Cotidiano* no. 50: 205–15.
Long, Norman. 1992a. "From Paradigm Lost to Paradigm Regained?" In *Battlefields of Knowledge: The Interlocking of Theory and Practice in Social Research and Development,* ed. Norman Long and Ann Long, pp. 16–43. London: Routledge.
——. 1992b. "Introduction." In *Battlefields of Knowledge: The Interlocking of Theory and Practice in Social Research and Development,* ed. Norman Long and Ann Long, pp. 3–15. London: Routledge.
Long, Norman, and Ann Long, eds. 1992. *Battlefields of Knowledge: The Interlocking of Theory and Practice in Social Research and Development.* London: Routledge.
McMichael, Philip, and David Myhre. 1991. "Global Regulation vs. the Nation-State: Agro-Food Systems and the New Politics of Capital." *Capital and Class* 43: 83–105.
Moguel, Julio, Carlota Botey, and Luis Hernández. 1992. *Autonomía y nuevos sujetos en el desarollo rural.* Mexico City: Siglo Veintiuno Editores.
Myhre, David. 1991. "Financing the Transformation of Mexican Agriculture: The Political Economy of Official Agricultural Credit in Mexico, 1976–1988." Thesis, Cornell University, Ithaca, N.Y.
Pessah, Raul. 1987. "Channeling Credit to the Mexican Countryside." In *The*

Search for Food Self-Sufficiency in Mexico, ed. James Austin and Gustavo Esteva, pp. 92–110. Ithaca: Cornell University Press.

Rello, Fernando. 1987. *State and Peasantry in Mexico: A Case Study of Rural Credit in La Laguna.* Geneva: United Nations Research Institute for Social Development.

Robles, Rosario. 1992. "La década perdida de la agricultura mexicana." *El Cotidiano* no. 50: 169–85.

Robles, Rosario, and Julio Moguel. 1990. "Agricultura y proyecto neoliberal." *El Cotidiano* no. 34: 3–12.

Sanderson, Steven E. 1986. *The Transformation of Mexican Agriculture: International Structure and the Politics of Rural Change.* Princeton: Princeton University Press.

Shwedel, Kenneth. 1992. "A Game of Wait and See: Agricultural Investment Slows in Aftermath of Ejido Reform." *Business Mexico* 2(12): 4–7.

PART III

GLOBAL REGION RESTRUCTURING

7

The Global Fresh Fruit and Vegetable System: An Industrial Organization Analysis

William H. Friedland

The fresh fruit and vegetable (FFV) industry is currently in the process of globalizing. This chapter undertakes an examination of this industry and studies several important firms that have become transnationalized.[1] My intent is twofold: first, to focus on the restructuring of an industry as it globalizes and, second, to understand the process by which firms, originally local, regional, or national in scope, become transnational in character.

The phenomenon of the multinational and transnational corporation has been studied since the 1960s. The behavior of these corporations has been analyzed to a considerable extent, although more in terms of the corporations' general effects on society and, in particular, Third World nations, than in terms of their organization and operation (Wimberley 1991). As private entities—moreover, as private entities spanning many nations—they rarely provide access to independent investigators because of their concerns that trade secrets might be revealed that would help their competitors.[2]

[1] This industry is also referred to as the "fresh produce" industry in this chapter. The FFV industry is a subsegment of a larger industry, fresh foods, which is in turn a subsector of the food industry. Although some segments of the food industry have been globalized for a long time, others have not. One that has not is fresh produce, which was characterized, except for one exceptional commodity (bananas), as being fragmented and primarily locally, regionally, or nationally based.

[2] There is a voluminous literature on "multinational," "transnational," and "global" corporations and enterprises (see, for example, Eels 1976, Ghertman and Allen 1984, and Taylor and Thrift 1986). On the history of multinationals and their development see Hertner and Jones 1986.

The basic argument of this essay is that, as the globalization of the FFV industry has begun, a handful of firms have emerged that have become transnational, although they are much smaller than firms in larger-scale industries such as automobiles, where not only the industry is globalized, but the various firms within the industry are themselves global in character, even if many of them still have a national base.[3]

In the production and handling of fresh fruits and vegetables from producer to consumer, I define three basic segments: producers, those who directly produce the product; marketers, those who handle the product to consumers (including retailers and food service); and distributors, those who serve as intermediaries between producers and marketers.[4]

With the increased production of fresh fruits and vegetables for long-distance and intercontinental distribution and marketing, it is the distribution segment that is globalizing. Firms in the production and marketing segments tend to remain locally, regionally, or nationally based, although the globalization of the distribution sector is being driven, in considerable part, by national concentration of food retailing in western Europe and a similar concentration, although on a regional basis, in North America.[5] This chapter focuses on the distribution segment.

Change and Its Causes in the FFV System[6]

With the urbanization that accompanied the industrial revolution, the provision of food for newly urban populations became a problem.

[3] In contrast to the automobile industry, there is the clothing industry, which is now thoroughly globalized as the casual inspection of "Made in . . ." labels in any clothing store in the world will reveal. The clothing industry, however, does not yet manifest tendencies by which manufacturing firms have become globalized. Thus, a distinction should be made between an industry being globalized and the various firms in that industry that may still be largely local or national in character.

[4] Others define the FFV system differently. R. Brian How defines the four major categories of marketing as "shipping point operations close to the area of production, long distance domestic transportation, wholesale operations at terminal or destination markets, and retailing or food service to the final consumer (1991:74–75). Max Brunk contends that the distinction is arbitrary: "We commonly make the distinction between production and marketing at any point. . . . Values added prior to any chosen point we commonly call 'marketing.' What is marketing to one person may be production to another" (1991:46).

[5] An interesting variation to this pattern occurred late in 1992 when a consortium of FFV producers in Mexico, backed by a Mexican bank, purchased Del Monte Tropical, an important transnational FFV corporation. Through this purchase, a group of primary producers integrated downstream into distribution.

[6] I have developed the analysis in this section in greater detail in Friedland 1994. See also Cook 1990.

Before the industrial revolution, most people produced their own food. As population aggregated in urban centers, the length of the working day and the unavailability of land precluded self-production of food; therefore, food had to be brought from rural agricultural settings to the new urban locations. These needs saw the rise of specialized production of foods such as sugar in the Caribbean (Mintz 1986) and grains such as wheat in eastern Europe. As urban populations grew, new production locations—California, the U.S. Midwest, Argentina, Canada, and Australia—entered into grain production for the export market. Thus, for the first time in world history a truly globalized food system began to emerge (Friedmann 1978a, 1978b, 1980), as did a handful of transnational firms, such as Cargill, Dreyfus, and Bunge, specializing in the distribution of grains (Morgan 1980).

Processing of foods grew in importance, and the technology of food preservation, particularly canning, improved. This stage of development saw the emergence of transnational food-processing firms such as Nestlé, Heinz, and Campbell. When frozen food technology got under way and the possibility of maintaining food in frozen condition from processor to consumer was in place, established processors and new firms moved into this form of food handling. Frozen foods tasted more like the original fresh foods than canned goods, and the canning industry withered.

The most recent stage in food consumption has witnessed an increase in the demand for fresh fruits and vegetables. Until the turn of the twentieth century, urban populations ate fresh produce only seasonally. The major change began with bananas, a tropical fruit that could withstand a long transportation link between producer and consumer if certain conditions could be obtained. The banana industry grew, and U.S. firms became extensively involved in the internal politics of Cuba and the Central American "banana republics." In the British, French, and Italian colonies banana production for the metropole gave rise to banana specialist firms in the metropole.

As metropolitan centers grew, seasonal production of fresh fruits and vegetables also developed. A distinctive group of distributors who served as a critical link between farm producers and markets began to handle this seasonal production. Initially scattered, the function of distribution became spatially concentrated in "terminal markets" where distributors, brokers, wholesalers, and retailers came together. Most distribution firms were very small, but the banana distributors—also involved in production—became important economically. Most distributors were involved with local markets; only a miniscule segment dealt with regional and national ones. It has been only since the 1980s, with the growth of consumption of fresh produce, that large firms have emerged that are involved with global production and distribution.

The production-distribution system of produce in the United States began to develop systematically during the 1920s. After that time, commodity after commodity began to be available on a year-round basis, lettuce being one of the early commodities (Friedland et al. 1981). Year-round lettuce availability was accomplished by coordinating production in different locations. Later, commodities such as tomatoes joined the year-round parade; tomatoes were grown in different locations, harvested when "mature green," and shipped to local ripeners who would heat the tomatoes and gas them with ethylene gas, a natural plant ripener. The fruit would turn red and look ripe.

Gradually, supermarket chains demanding a steady and predictable flow of fresh produce became prevalent. Since the early 1980s two major developments have created a fundamental change in the advanced industrial countries, where most people now expect to have a wide variety of fruits and vegetables available on a year-round basis. The first has been the extension of the production season through plant-breeding programs, changes in horticultural practices, and the development of many production locations. The second has been the expansion of varieties of fruits and vegetables, particularly tropicals.

During the 1980s a host of tropical exotic fruits and vegetables were introduced on the international market. Consumers began buying fruits such as atemoya, breadfruit, cherimoya (custard apple), carambola (starfruit), feijoa (pineapple guava), guava, kiwano (horned melon), lychee, passionfruit, and sapote and vegetables such as bok choy, cassava (yuca), chayote, daikon (oriental radish), fava beans, fennel, jicama, nopales (cactus), and plantain. Also new to consumers were many salad greens, such as arugula, radicchio, chicory, oak leaf, and, an even newer development, baby vegetables. Many of these new exotics were originally intended for specialized markets. For example, jicama for the Mexican market in California and the Southwest and plantains for consumers from the Caribbean. But they are now being purchased in considerable volume by consumers in Western Europe and the United States.

These developments have also been driven by the combination of (1) the health-food movement and concerns about food safety, (2) the development of production capacity through the export of capital and technological expertise, and (3) the establishment of capital-intensive "cool chains," which maintain chilled temperatures from origin to consumer.

The distances involved, which stretch over thousands of miles and across continents, have created a transnational industry in which products are grown on five continents and sourced around the world for distribution and sale on three continents (North America, Europe, and Japan and Hong Kong in Asia).[7]

The international growth of the fresh produce industry historically has been very uneven. Experiments in the banana trade began as early as the 1870s (CEPAL 1979: 114). But the early trade, though *inter*national, was not *trans*national in that nationally based firms produced bananas in tropical colonies (or semicolonies in the case of the United States) for consumption in the metropole. The trade in other fresh fruits and vegetables did not become truly transnational until sourcing of production spread widely beyond national borders and a nearly global market developed.

But sourcing is still very uneven. Although South America, Africa, and Asia provide fresh produce for the North American, European, and Japanese/Hong Kong markets, only a few countries on these continents are significant sources—Chile, Argentina, Brazil, and Uruguay in South America; Kenya, Zimbabwe, and South Africa in Africa; and Malaysia and Thailand in Asia. Some countries in South America, such as Peru and Bolivia, are better known for other horticultural exports (e.g., cocaine), and Columbia is a major flower exporter but contributes little to the global fresh produce system.

Consumption of fresh produce via the global system is restricted to the United States and Canada in North America, the Western European countries and Scandinavia in Europe, Japan and Hong Kong in Asia, and the Arab oil states in the Middle East. Fresh produce is consumed worldwide, but in countries, with large populations, such as China, India, and Nigeria, there is only local production for local consumption.

Industrial Organization

Three distinct elements characterize the fresh fruit and vegetable industry. First, the industry consists of three separate segments, only one

[7] Despite the enormous expansion of the FFV industry, much of production, distribution, and marketing still remains national (in the case of the USA) and intraregional (in the case of Western Europe). Much of the new global trade is south-north, with production taking place in the lesser developed south for consumption in the advanced capitalist north. Southern agri-

of which, the distribution segment, is truly transnationalized. The other two segments, production and marketing, though showing a few tendencies toward transnationalization, lean more toward a localized, regional, or national character.

Second, the distribution segment of the industry is probably highly concentrated, even if, at the moment, there are few data available on the degree of concentration. The trade literature makes clear that only a handful of firms have established a global presence. That presence is demonstrated through (1) advertising and news reports in trade publications, (2) news reports on financial pages, and (3) word-of-mouth comments by individuals involved in the industry. There are, in addition, a significant number of "wannabe" firms (i.e., that want to be big and transnational). I have identified only a few firms as major players in the industry, but there are probably another dozen wannabe firms.

Third, there is significant variation in history, size, and internal structure of the firms involved in FFV distribution. One major characteristic differentiates the four or five largest firms—whether they had a banana base historically or are new firms, either without a banana base or in the process of developing one.

In an initial analysis of industrial organization, certain variables are important in understanding the character of the major players in distribution. These variables will be applied to the largest firms: Dole, Chiquita, Albert Fisher, Polly Peck, and Del Monte. Dole is a U.S.-based transnational known until 1991 as Castle and Cook. It originally began as a merchant firm in the Hawaiian Islands and subsequently became involved in food processing, real estate, and FFV activities. Chiquita is also originally U.S.-based and formerly was known as United Brands, and before that as the United Fruit Company. Albert Fisher is a British-based firm with extensive holdings in North America (particularly Florida and California), Great Britain, and the European Community (especially the Netherlands and Germany). Polly Peck International, a British-based transnational conglomerate, is currently in bankruptcy proceedings and being dismantled. The Del Monte Fresh Produce Company, originally a U.S.-based company, was known as Del Monte Tropical Products until late 1992.[8]

cultural export expansion is often associated with national debt-servicing measures or restructuring the agricultural sector to make it more profit-accumulative.

[8] I shall refer to Del Monte Fresh Produce Company as Del Monte Tropical Products (DMT). DMT has had a very complex history. The firm originated as part of the original Del Monte Corporation, a major food processor, when it diversified 1962. The Del Monte Corporation moved into bananas because it feared a takeover by the United Fruit Company. Because of an agreement between United and the U.S. government in 1958, the purchase by

Of these firms, Dole, Chiquita, and Del Monte Tropical Products (DMT) have histories in banana production. In contrast, Albert Fisher and Polly Peck International are new firms, each with lengthy individual histories predating their rebirth as global FFV firms. Neither had any background in banana production or distribution. Bought by Polly Peck in 1989, DMT brought to its new owner its long-standing banana operations as well as an aggressive expansion program in several new fruit and vegetable commodities.

Structure

The FFV industry—production, distribution, and marketing—can be conceptualized in the shape of a dumbbell, with production and marketing being the two large weights on each end connected by the narrow channel of distribution.

This metaphor conveys several ideas. First, production and marketing are much larger than distribution in terms of the numbers of people involved. Distribution, in contrast, is very capital- and energy-intensive, requiring trucks, airplanes, and ships, all with refrigeration capacity. It is also the system that deals with physically distant spaces, transporting fresh fruits and vegetables between continents over thousands of miles.

A somewhat different diagram might be drawn if value added were used instead of numbers of workers involved. In the United States, for example, if we use the value-adding data provided by R. Brian How (1991:79), what is referred to as the distribution segment becomes far

---

the Del Monte Corporation of the West Indies Fruit Company, a banana producer, precluded the takeover by United (Braznell 1982:149–50). From then on the Del Monte Corporation aggressively expanded its banana holdings, creating a subsidiary firm, DMT, as it ventured into other fresh fruits and vegetables.

The Del Monte Corporation and its subsidiary DMT were acquired by RJR in 1979, the former R. J. Reynolds Tobacco Company, as that company diversified its holdings out of tobacco into food. In 1989 RJR was in turn taken over by Kohlberg, Kravis, and Roberts (KKR), in one of the largest corporate takeovers at the time (Burrough and Helyar 1991). The KKR takeover, involving the acceptance of a staggering debt load, required KKR to sell off pieces of the RJR empire. During 1989 KKR first sold DMT to Polly Peck International for $875 million (*Eurofruit* [London], October 1989:1, 3); later it sold the food processor Del Monte Corporation to a consortium including Merrill Lynch, Citicorp, Kikkoman, and a clutch of Del Monte managers (Lehrman 1989). The acquisition of DMT converted Polly Peck into a genuinely global corporation in FFV, a transition that lasted until Polly Peck went bankrupt in 1990. As part of the bankruptcy liquidation, in 1992 Polly Peck's administrators sold DMT to a consortium of Mexican investors, led by Grupo Cabal, backed by a Mexican bank, Nacional Financiera, with a 60 percent interest, for $499 million (*Eurofruit,* October 1992:5). The separation of DMT from Polly Peck took place in 1992 and has been too recent to be able to report on each separate firm; accordingly, this essay will deal with DMT as of the period when it was a subsidiary of Polly Peck.

more important. Thus, How values production (domestic and imported) at US$12.2 billion and what I refer to as marketing at $18.41 billion. How's three categories of shipping-point operations, long-distance transport, and wholesaling, which are approximately equivalent to my category of distribution, are valued at $13.47 billion. Of the total value-added activity, production accounts for 27.7 percent, marketing for 41.8 percent, and distribution for 30.6 percent.

The dumbbell metaphor also illustrates the articulation between the three segments. Most distributors are involved in some degree with production, but impinge on marketing only to a limited degree. This impingement occurs primarily through brand labeling, the practice of placing labels on produce which identify the product as "Chiquita" or "Del Monte" or "Dole." A few producers have integrated downstream into some aspects of distribution but never so far forward as to impinge on marketing. Marketers are specialized in their area, and I know of no case of their involvement in either production or distribution. What is also the case is that historical distribution firms—those with banana histories—have long been involved in direct production of this commodity but usually have become only partially involved in production of other commodities, that is, the new commodities now in year-round production for a global market.

Partial involvement in production by distribution firms refers to myriad practices by which these firms source their commodities. These can involve practices such as contracting and joint ventures of various types. In contracting, distribution firms contract with production firms, cooperatives, or individuals for delivery of commodities, usually with specifications as to timing, quality of product, and price. In joint ventures, distribution firms may engage in practices in which all or part of the capital requirements may be provided by the distribution firm, with specifications on some aspects of production, quality, and the division of income from the sale of the commodities.

There are also many kinds of brokers, jobbers, agents, and others who undertake liaison responsibilities in the distribution chain. Most of these agents are local, although some operate at the national level.

It is with the larger distribution firms, however, that the process of transnationalization is occurring. A handful of these firms have risen above their original national bases and consciously are developing strategies to expand their sources of production and their markets.

Table 7.1 sets out in summary form the major factors used in my analysis of firms.[9] These factors include the following:

[9] In Table 7.1 I used the data available on Polly Peck after it had acquired DMT in 1989. It was not feasible to provide separate data on DMT, which was an integral part of the Del

- *Base*—which consists of two factors: the national base and whether the firm has a historical commodity base
- *Number of countries*—as represented by the number of countries in which the firm or its subsidiaries are present, not including all countries in which the firm does business
- *Number of subsidiaries*—as represented by wholly or partially owned subsidiaries
- *Conglomerate*—whether the firm has substantial financial interests in nonfood or nonfresh produce activity or both
- *Vertical integration*—the degree to which the firm integrates upstream into production (growing), is engaged in transportation and if so in what form, and is oriented toward the retailing segment of the market
- *Value adding*—whether the firm, as a matter of policy or through its behavior, engages in adding value to produce, in other words, engages in activities that process commodities to some degree or otherwise enhance the value of commodities in the market

Transnational Firm Organization[10]

A discussion of the major transnational players in the FFV distribution system should begin by making a distinction between three older firms—Chiquita, Dole, and Del Monte Tropical—and two newer firms essentially initiated in the early 1980s—the Albert Fisher Group and Polly Peck International.[11] The first three have histories in banana production and distribution—Chiquita being the oldest—whereas the two newcomers have no banana histories.

The significance of banana history becomes clear when it is understood that the older firms expanded beyond their original commodity base (bananas for Chiquita and Del Monte, bananas and pineapples for Dole) into other fresh fruits and vegetables. Chiquita began considering movement into other commodities during the 1960s, reasoning that its expertise and distribution system in bananas could be used for other commodities. In 1969 Chiquita bought seven lettuce-producing firms in California and integrated them into a single subsidiary, Interharvest, instantaneously making it the largest single firm in U.S. lettuce produc-

Monte Corporation when the latter was acquired by RJR in 1979. The structural components of DMT have never been set out publicly to my knowledge.

[10] This section has been derived from CEPAL 1979, corporation reports from Standard and Poor, and from the *New York Times*.

[11] The peculiar status of Polly Peck/DMT poses problems for this discussion. When it acquired DMT in 1989 for $875 million, Polly Peck was already of major importance globally in FFV, although its fresh produce activity was not as important as that of DMT. But DMT was never integrated into Polly Peck operations. With the sale of DMT to a Mexican consortium in 1992 and Polly Peck in receivership, a differentiated analysis would be useful, but such data are not yet available. I therefore provide data on Polly Peck/DMT before the breakup.

Table 7.1. Major fresh fruit and vegetables corporations

	Chiquita	Dole/Castle and Cook	Polly Peck/DMT	Albert Fisher
National base	USA	USA	UK, North Cyprus, Turkey, USA	U.K.
Commodity base	Bananas	Pineapples, bananas	Citrus (Polly Peck), bananas (DMT)	None
No. of countries[a]	14+	15	10+	10
No. of subsidiaries	240+	37	45	71
Conglomerate	No	Yes; Hawaii real estate	Yes; electronics, recreation	No
Vertical integration				
Growing	Yes	Yes	Yes	No
Transport	Ships	Ships	Ships, air	Minor road
Retail	Labeling	Labeling	Labeling (DMT)	None
Value adding	Yes	Yes	Yes	Yes

Sources: *Moody's Industrial Manual,* vol. 1, 1990 (New York: Moody's Investors Services); *Who Owns Whom 1990: United Kingdom and Ireland,* vol. 1 (London: Dun & Bradstreet); *The Packer, Eurofruit.*

[a]Represents the number of countries within which the firm has subsidaries, *not* the number of countries within which the firm distributes.

tion and distribution. This venture proved unsuccessful, and Chiquita closed down Interharvest in 1983 (*San Jose Mercury News,* August 13, 1983, sec. B, p. 1). Chiquita moved back into FFV later in the decade when it saw the successes of its competitors, Dole and DMT.

The Dole Food Company emerged in its modern form in 1961 when Castle and Cooke acquired the Dole Company, a pineapple producer. It expanded into bananas in 1964 and 1967 as it bought an increasing share of the Standard Fruit Company, a banana producer for the North American market. In 1967, when Dole owned 87 percent of Standard, Standard supplied 31 percent of North American banana requirements (CEPAL 1979:80). By 1977, when United Brands (now Chiquita Brands International) was becoming disenchanted with its lettuce venture, Dole purchased Bud Antle, the second largest lettuce producer in the United States. It was from the Bud Antle base that Dole expanded into a wide variety of other commodities.

During the 1980s, Chiquita, Del Monte, and Dole expanded substantially into global sourcing and distribution based on their banana operations. At the same time two new firms without a banana base established themselves in FFV. Polly Peck began as a small firm in the British "rag trade," making clothing. Taken over in 1980 by Asil Nadir, a Turkish Cypriot whose family had moved to the United Kingdom after the Turkish invasion of Cyprus, the firm was converted into Polly Peck International, taking advantage of citrus orchards in the part of Cyprus occupied by the Turks (Hindle 1991:chap. 3).

Through close connections with the Northern Cyprus government, unrecognized by all states except Turkey, Nadir opened a box factory to ship citrus to Europe. This trade proved so lucrative that Polly Peck stock values rose continuously. Nadir was able to raise substantial capital resources in Britain and Turkey and by 1989 had expanded into electronics and recreation. He expanded citrus operations to Turkey, and Polly Peck became a major distributor of fresh fruits and vegetables to Europe and the Middle East. The acquisition in 1989 of DMT and Sansui, the Japanese electronics firm, moved Polly Peck into the top one hundred British corporations.

Polly Peck International's startling growth during the 1980s was unlike the experience of Chiquita, Del Monte, and Dole, which had expanded from their banana bases. Polly Peck, in contrast, developed a hitherto undeveloped citrus base and found it so profitable that this base became the driving engine for the emergence of this new transnational corporation. All came tumbling down in 1990 when the debt load of Polly Peck, compounded by the British equivalent of insider trading, led to the collapse of the company. DMT in the meantime had been "swallowed" by Polly Peck but not "digested." With the sale of DMT to Mexican interests in 1992, a whole new phase of the firm's development had begun, but that experience is too new to have accumulated any history.

The Albert Fisher story is different from that of the banana-based firms or Polly Peck. The Albert Fisher Group (AFG) represents an interesting case in that the firm manifests distinct transnational characteristics while at the same time being focused on a relatively small number of national markets.

Albert Fisher was originally a sizable, yet small, produce operation in Great Britain, until it was bought by entrepreneur Tony Millar in 1982. Millar turned Albert Fisher into the Albert Fisher Group, an important FFV transnational with major holdings in the United Kingdom the United States, the Netherlands, Germany, and several other countries.

How does one take a minor FFV distributor and turn it into a transnational enterprise in less than a decade? First, to make the initial purchases, one must bring in capital, one's own and that of other investors, especially institutional investors. Millar brought enough capital with him from his previous financial involvements to buy a sizable personal holdng of AFG's shares at the very outset. But other capital is required, and Millar amassed this through his previous experience and contacts, with the result that a significant number of institutional shareholders supported him in the AFG expansion.

Second, new capital must be continuously generated so that the firm grows through acquisition. To accomplish this, management must demonstrate that profit levels will improve *organically*. Under such conditions, investors will buy new issues of stock as they are required for growth. In Fisher's case, new capital was generated by regularly offering new stock on a one-for-three basis with share prices set below the current value in the market.

The third critical element is to make good acquisitions, perhaps the most delicate part of the operation, since target firms must have established earnings patterns *and* the potential for improvement. In addition, new acquisitions have to fit into an overall strategy. For Fisher, this strategy involved three processes: (1) clustering of acquisitions geographically, rather than spreading all over the world, and consolidating acquisitions before venturing into new areas; (2) choosing firms focused on providing FFV to food-service companies; and (3) choosing firms with good managements that were then locked into place in short-, medium-, and long-term contracts.

Millar accomplished the first of these strategies by acquiring family-based firms (with the exception of three firms in a single package) in distinctive but strategic locations. The second strategy was accomplished by acquiring firms that mainly catered to food-service providers rather than firms selling to supermarkets or greengrocers. To achieve the third strategy, Millar acquired (still with the single exception noted above) firms with some cash, some shares in AFG, and a deferred payment plan conditioned on profits during the next two to three years. The managements of the acquired firms were always (with that one exception) locked in with two- to three-year management contracts *and* with various forms of profit sharing. AFG bought clusters of firms in Florida and California so that, by 1988, AFG controlled 20 percent of Florida distribution and 12 percent of California's much larger distribution capacity (*Financial Weekly* [London], August 11, 1988). Fisher also has a major stake in the Netherlands and Germany, as well as growing importance in Belgium and France.

Having established significant bases in major distribution locations, AFG has institutionalized a number of important management practices. First is the exchange of management information. When Fisher companies have a problem, managers of other Fisher companies are brought in to provide advice. Second is consolidated purchasing. With so many companies buying commodities, central purchasing permits AFG to wield greater market power and achieve economies of scale. AFG's main office is in the United Kingdom, where it maintains a tiny headquarters staff of twenty-two people (*The Independent on Sunday* [London], October 14, 1990:10).

In its U.S. operations, AFG focuses on a particular market, food service. Although AFG's companies also sell fresh produce to wholesalers, it concentrates on firms that transform raw food into foods eaten in restaurants, institutions, cruise ships, and so on. With Americans now eating approximately half their meals outside the home, AFG recognizes food service to be the dynamic segment of the FFV industry. In developing its acquisitions strategy, AFG acquired firms with histories of sales to food-service companies.

The Albert Fisher Group is not a conglomerate. Unlike Chiquita, Del Monte, Dole, and Polly Peck, all of which have major stakes in transportation (particularly in refrigerated cargo ships), AFG has only some minor road transport capability in the United Kingdom. AFG has an aggressive policy toward value adding which has meant that it has acquired significant capacity in firms that process and prepare foods. Thus, some of its subsidiaries are involved with fish, shellfish, pickles, and other bottled foods. It has not integrated upstream toward growing or downstream into retailing, but its policy is to get as close to each end of the food chain as possible. Fisher refuses to become involved with food labeling, unlike Dole, Chiquita, and Del Monte, thereby distancing itself from retailing.

Like Polly Peck, AFG is a product of the expansion of the FFV industry in the 1980s. Both firms made notable marks on the London stock market. Fisher's Tony Millar, however, fits better with the British financial establishment because, unlike Asil Nadir, he has been modest in seeking press coverage, is himself a native-born Englishman, and has walked carefully through the minefields of the City (London's Financial district). In ten years as head of AFG, Millar had institutional stockholders muttering about his procedures only once and, in that case, he backed off in a hurry and continued to keep them happy. This prudent march did not preserve Millar's leadership of AFG once the firm's stock values declined in 1992 because of the world recession, and he had to resign as the head of the Albert Fisher Group (*Eurofruit,* September 1992:4).

Conclusions

The differences between the FFV industry and other transnationalized industries must be acknowledged. In contrast to the larger industrial groups, such as the automobile industry, where capital investments involve hundreds of billions of U.S. dollars, the FFV industry is a comparatively modest undertaking valued only in the hundreds of millions of dollars. Nor is the industry as broadly transnationalized as the automobile industry and others, where sourcing takes place in dozens of countries, manufacturing in a considerable number of countries, and marketing in every country of the world. The FFV industry, however, provides an opportunity to study the process of transnationalization and some of its important firms to understand the kinds of strategies necessary to survive and grow as the industry restructures itself toward transnationalization.

First, the FFV industry is transnationalizing only in its distributional segment, not in its production or marketing segments. Distribution is the vital link between production sources and wholesale and retail markets. It involves firms that have developed the expertise and capital capabilities to act as agents between producers willing to produce for markets they do not understand and marketers who know the character of their market but cannot handle the logistics of delivery of highly perishable commodities. Although it is the marketers who drive the system with their demands for quality, volume, price, and predictibility, it is the distributors who assemble capital and logistics to deliver perishable commodities in salable form.

The assemblage of transportation logistics and knowledge of markets is complex. As producers who have waited unsuccessfully for transport of their product in countries such as Zimbabwe know, finding appropriate chilled containers with appropriate air transport at exactly the right moment makes the difference between profit and loss (*Eurofruit,* August 1990:19–22). The capability—in terms of skills and capital resources—to accomplish this makes the role of the distribution firms critical in the whole process, while at the same time they remain secondary players to the dominant retailers, particularly the buyers for the supermarket chains.

For the firms involved in distribution, two basic types of strategies to meet the new transnationalized markets for FFV have been followed. The first has built from an established commodity base, primarily bananas, into other commodities, most frequently through acquisitions. The second approach involves assembling capital resources to acquire significant numbers of small firms (Albert Fisher) or important firms (Polly Peck) with established records in importing or distributing fresh produce from a variety of locations to provide broad geographical and commodity availability. Both strategies are based on the growth of consumption and the ability to source commodities, to create new or utilize existing transportation systems, and to deliver commodities in volume to marketers.

What is less clear from this account, and requires additional research, are the methods of internal organization used by the firms to implement their strategies. Tentatively, it is clear that some firms, Chiquita, Dole, and Del Monte Tropical being examples, source tropical products because of well-established banana operations that are extended into new commodities. Other firms, such as Albert Fisher, develop sourcing by the purchase of small firms with some sourcing capability. Through the assemblage of the many, the parent firm develops a wide range of sourcing capabilities. The parent firm must then engender a degree of integration between its many subsidiaries. The degree to which Albert Fisher has accomplished this internal coordination is unclear; the firm's pattern of retaining the former owners as managers works well to maintain high returns, but it is less obvious that these various enterpreneurs know how to work together or that Fisher's central management has devised the means to capture the full benefits of economies of scale.

Banana-based firms, such as Chiquita, Del Monte, and Dole, have considerable experience in the development of coherent internal organization. Their banana experiences produced very highly integrated and centralized internal structures. Yet the extension of such structures to other commodities is not a trivial matter, as Chiquita's earlier experience in the formation of Interharvest proved. Chiquita appears to be exploring some form of integration in its acquisition of a substantial holding in Pascual Hermanos, one of the larger Spanish grower-shipper firms (*Eurofruit,* July 1991:3; October 1991:2). In this respect, Dole and Del Monte appear to have been able to extend beyond their banana base with greater facility than Chiquita. Why this is so can become clear only when studies have been made comparatively of internal structures and acquisition strategies, at present a difficult task.

Despite the methodological and other difficulties, the analysis of an industrial sector in the process of restructuring and globalizing remains an interesting task. The literature thus far, although having opened an entire new arena to understanding, needs considerable broadening if the process of transnationalization—and its societal consequences (Borrego 1981)—are to be understood.

References

Borrego, John. 1981. "Metanational Capitalist Accumulation and the Emerging Paradigm of Revolutionist Accumulation." *Review* 4(4): 713–77.

Braznell, William. 1982. *California's Finest: The History of Del Monte Corporation and the Del Monte Brand.* San Francisco: Del Monte Corporation.

Brunk, Max. 1991. "Agricultural Marketing: Finding Opportunities." *Produce Business* 7(12): 46.

Burrough, Bryan, and John Helyar. 1991. *Barbarians at the Gate: The Fall of RJR Nabisco.* New York: Harper Perennial.

CEPAL (United Nations Economic Commission for Latin America). 1979. *Transnational Corporations in the Banana Industry of Central America.* New York: United Nations Economic and Social Council.

Cook, Roberta L. 1991. "Challenges and Opportunities in the U.S. Fresh Produce Industry." *Journal of Food Distribution Research* 21(1): 67–74.

Eells, Richard. 1976. *Global Corporations: The Emerging System of World Economic Power.* New York: Free Press.

Friedland, William H. 1994. "The New Globalization: The Case of Fresh Produce." In *From Columbus to Conagra: The Globalization of Food and Agriculture,* ed. Alessandro Bonnano et al. Lawrence: University Press of Kansas.

Friedland, William H., Amy E. Barton, and Robert J. Thomas. 1981. *Manufacturing Green Gold: Capital, Labor, and Technology in the Lettuce Industry.* New York: Cambridge University Press.

Friedmann, Harriet. 1978a. "Simple Commodity Production and Wage Labour in the American Plains." *Journal of Peasant Studies* 6(1): 71–100.

——. 1978b. "World Market, State, and Family Farm: Social Bases of Household Production in an Era of Wage Labour." *Comparative Studies in Society and History* 20(4): 545–86.

——. 1980. "Household Production and the National Economy: Concepts for the Analysis of Agrarian Formations." *Journal of Peasant Studies* 7(2): 158–84.

Ghertman, Michel, and Margaret Allen. 1984. *An Introduction to the Multinationals.* London: Macmillan.

Hertner, Peter, and Geoffrey Jones, eds. 1986. *Multinationals: Theory and History.* London: Gower.

Hindle, Tim. 1991. *The Sultan of Berkeley Square: Asil Nadir and the Thatcher Years.* London: Macmillan.

How, R. Brian. 1991. *Marketing Fresh Fruits and Vegetables.* New York: Van Nostrand Reinhold.

Lehrman, Sally. 1989. "Preserving the Good Name: Reassurances Accompany the Sale of Del Monte." *California Farmer* 271(8): 6–7.

Mintz, Sidney W. 1986. *Sweetness and Power: The Place of Sugar in Modern History.* New York: Viking.

Morgan, Dan. 1980. *Merchants of Grain.* New York: Penguin.

Taylor, Michael, and Nigel Thrift, eds. 1986. *Multinationals and the Restructuring of the World Economy.* London: Croom Helm.

Wimberley, Dale W. 1991. "Transnational Corporate Investment and Food Consumption in the Third World: A Cross-National Comparison." *Rural Sociology* 56(3): 406–31.

8

Comparative Advantages and Disadvantages in Latin American Nontraditional Fruit and Vegetable Exports

Luis Llambi

During the 1970s and 1980s, some Latin American countries became important exporters of fruits and vegetables for specialty markets in affluent countries. Chile is the major off-season exporter of fresh fruits during the Northern Hemisphere winter months. Table grapes, apples, and pears alone account for roughly 90 percent of total export volume, although Chile is diversifying the produce exported. Apricot, cherry, peach, nectarine, prune, berry, kiwifruit, and avocado exports are rapidly increasing as additional orchards reach mature-yield levels.

Brazil is the world's largest orange juice producer and exporter, accounting for nearly 80 percent of world shipments. Frozen concentrated orange juice is Brazil's third most important agricultural export after coffee and soybean products.

In 1990 the United States imported from Mexico more than $1.6 billion worth of vegetables, both fresh and processed. Mexico provides about half of the fresh tomatoes consumed in the United States during the winter months, sharing the market with Florida growers. Following tomatoes, in value terms, were bell peppers, cucumbers, onions, cantaloupes and squash. Production and export of "temperate" crops such as asparagus, broccoli, brussels sprouts, cauliflower, and celery are increasing. Also growing are Mexico's exports to the United States of

I am indebted to Philip McMichael, Laura Raynolds, and Enrique Figueroa for their valuable suggestions and assistance in writing this essay.

both fresh and processed fruits, including strawberries, mangoes, and citrus (see Foreign Agricultural Service, May 1991).

Colombia is the second largest world exporter of fresh cut flowers after Holland. Flowers currently are Colombia's third most important agricultural export after coffee and bananas. About 80 percent went to the U.S. market, particularly during the winter season. Other Latin American suppliers of cut flowers to the United States include Costa Rica, Mexico, Ecuador, and Guatemala.

Since 1983, when the United States provided duty-free treatment to some selected exports from twenty-four Central American and Caribbean countries—the so-called Caribbean Basin Initiative (CBI)—Costa Rica, the Dominican Republic, and Guatemala have become the main U.S. trading partners in the area. In 1990 fresh pineapples were the most important horticultural CBI export item, other than bananas and plantains. Fresh melons, citrus juice, and frozen vegetables followed in rank (Ballenger 1991; White 1991; Lopez et al. 1989).

These are the success stories, but there have been some failures. As new export opportunities are appearing, others are fading, and as more growers and countries are trying to enter these markets, it is important to assess what their prospects are and what policies might be designed to shape their process.

This essay has three main objectives. The first is to explain both the takeoff of nontraditional fruit and vegetable exports in particular Latin American countries since the 1970s and the economic performance of these countries as exporters since then.

The second goal is to assess the contribution of these nontraditional exports to Latin American economic growth. Fruit and vegetable exports have been depicted as a promising substitute for the traditional exports of the region, since the demand for these latter commodities is either stagnating or growing only slowly (see von Braun, Hotchkiss, and Immink 1989). National governments, bilateral agencies such as the U.S. Agency for International Development, and multilateral institutions such as the World Bank are according a high priority to these exports as a sustainable solution for stagnant Latin American agricultural sectors and overwhelming financial constraints. Taking into account not just a short-term and regionally localized perspective, but also a global and long-term view, might help us to anticipate their real future opportunities and biggest constraints.

The third objective is to identify some of the relevant policy issues that must be considered in designing new policy strategies in which nontraditional fruit and vegetable exports might find a place.

To tackle these issues I depart here from standard international trade theories focusing on market-led comparative advantages to explain exchange flows across borders. Instead, a political economy perspective is developed that focuses on international competitiveness as the result of the strategies adopted by national governments, transnational capitals, and so-called multilateral (i.e., global) institutions. Each of these agents tries to shape the conditions in which trade takes place.

To explain their outcomes and to forecast future trends, we need to scrutinize these agents' strategies, beginning with private capitals directly participating in the commodity circuits. Since most fruit and vegetable exports are capital-intensive and demand large investments and coordination, at both their productive and marketing stages, transnational corporations have been gaining ascendance as the main actors in these circuits. Being transnational means not only that these firms market their products globally, but also that their investment allocations may move relatively unhindered between different locations of the globe in search of enhanced business opportunities. Transnational corporations are therefore in a very good position not only to reap each location's comparative advantages but also to overcome the constraints imposed by national governments' regulations (see Bonanno 1991, Raynolds 1991).

The enhanced bargaining power of transnational firms vis-à-vis national governments should not lead us to underestimate the power that still remains in the hands of particular nation-states. Even Third World governments, weakened by the deregulation policies imposed on them as a result of the debt crisis, still retain some powerful instruments to influence market forces in their behalf. Not only sectoral- and commodity-specific policies but also broad macroeconomic maneuvers shape each country's options for enhancing its competitiveness in a particular international market. In an increasingly interdependent world national policies intersect. Yet, given the basic asymmetry of the global political order, major players in each particular market have more potential to affect other countries' performances through their own national policies.

Short-term comparative advantages can explain the emergence of paticular commodities in some regions, countries, or localities. Natural conditions and relative prices play a role in the competitive edge that some countries may enjoy in the export of some commodities at any given time. International competitiveness is not, however, the result of unhindered market competition, but one outcome of the relations between actions taken by various political entities—including national governments—to shape trade flows.

In a long-run perspective, the export possibilities for a particular country or region of the world depend on who decides the rules of the game in which market competition takes place. In the foreseeable future the prospects for Latin America's export possibilities will depend to a large extent on the ultimate balance between the trends toward increased market liberalization and globalization—the GATT negotiations being just one of its visible manifestations—and the trends toward formation of trade blocs. Since there is no longer a hegemonic power in the game, the possible outcomes will be determined in large part by the bargaining power achieved by the three main players (the USA, the EC, and Japan) through their ability to forge alliances with the other minor actors in the system.[1]

Explaining Latin America's Nontraditional Fruit and Vegetable Exports

The emergence of nontraditional export opportunities in the fruit and vegetable industries since the 1970s in Latin America was mainly the result of the following trends: the deteriorating market conditions of the Third World's traditional exports during the 1970s and 1980s, the First World's barriers to entry of most of Latin America's manufactured products, and the structural adjustment policies imposed by the IMF and the World Bank since 1982 on Third World debtor countries as a consequence of the increased leverage the debt crisis provided the transnational financial community to regulate international markets. Table 8.1 shows the shifting market shares in nontraditional agricultural exports held by several Latin American countries.

Latin American nontraditional fruit and vegetable exports are thus emerging and evolving in the midst of a transition from the inward-oriented strategies adopted in the region since the mid-1940s to the outward-oriented cum free market growth strategies that are now the

[1] It is also true that in the short run particular contingencies may explain some cyclical fluctuations in the international markets. Several bad or good weather seasons in a row, a shift in the economic policies of both exporter and importer countries, and other natural and political contingencies may explain some of the short-run cycles in the international markets of particular commodities. Yet these short-term fluctuations tend to be subsumed, and therefore have to be explained, within the broader long-term trends of the evolving economic and political order.

Table 8.1. Market shares (%) of selected Latin American nontraditional agricultural exports

	1978	1979	1980	1981	1982	1983	1984	1985	1986	1987	1988	1989	1990
Chile													
Grapes	6.7	7.0	7.9	11.2	15.8	17.1	23.7	27.1	24.8	24.1	24.9	23.7	22.9
Mexico													
Tomatoes, fresh	23.8	20.3	16.8	21.9	15.4	18.9	21.1	20.7	28.6	13.6	15.3	13.2	21.0
Dominican Republic													
Pineapples, fresh	0.02	0.04	0.002	NA	1.5	1.8	1.7	0.9	1.4	1.3	1.5	2.6	3.9
Honduras													
Pineapples, fresh	10.3	8.9	7.0	NA	10.5	11.5	13.8	10.4	6.2	7.4	7.4	6.1	8.7
Costa Rica													
Pineapples, fresh	NA	NA	NA	NA	NA	NA	NA	5.2	9.0	11.0	11.0	23.0	20.1
Brazil													
Orange juice	53.8	45.9	58.2	62.6	57.0	58.3	73.9	67.5	71.2	48.8	64.4	NA	NA
Colombia													
Cut flowers	5.6	7.4	8.4	10.3	10.5	10.8	11.7	11.6	12.8	7.3	7.6	NA	NA

Source: Brazil and Colombia, UN International Trade Statistic Yearbooks 1978–88; other countries, FAO Trade Yearbooks, 1978–90.
Note: NA, not available.

rule. Within this context nontraditional horticultural exports are emerging mainly as a substitute for declining traditional exports. Traditional tropical exports, in particular, are declining in importance because of substantial oversupply and stagnant (or only slowly growing) demand (Islam 1990).

Not only are First World policies dampening Third World traditional market opportunities but, despite GATT's success in lowering tariffs for manufactured products, nontariff barriers have increasingly been erected to restrain the Third World's attempts to penetrate labor-intensive high value-added industrial sectors: textiles, automobiles, processed foods, and so on (Gilpin 1987). Tariff escalation and nontariff barriers in processed fruits and vegetables markets are just such a case in point. According to Murray Cobban: "There are duties on imports into developed countries of almost all semi-processed tropical products except tea, most essential oils and jute, and on all processed tropical products" (1988:246).

Emergence and Location: Comparative Perspectives

The emergence and location of particular export-oriented investment opportunities are more adequately conceived as the result of international market condition (the demand side), each country's comparative advantages (the supply side), and state intervention.[2]

International Market Conditions: The Demand Side. A significant market size and a steady increase in demand are two main prerequisites for increased export opportunities. The dynamism of "fresh" and "processed" fruit and vegetable markets in recent years has been based largely on the changing lifestyles and demographic trends taking place in the industrialized world. A "shift from diets previously based on durable foods and meats" (Friedland 1991: 48) and the emergence of "clearly defined up-scale and down-scale markets" (Cook 1990:69) are the main trends governing these markets.

Nonetheless, except for the massive markets (bananas, tomatoes, citrus juices), the international sales of most products, when considered one by one, are only a few million U.S. dollars per year. Consequently,

[2] Focusing on the limits and possibilities of comparative advantages in shaping current trade trends, I stress the dynamic interplay between international market conditions and national-state interventions. The roles played by transnational corporations, the main actors at both the productive and marketing stages in most global fruit and vegetable circuits, are significant. Available data, however, are sparse (as also suggested in Chapter 7).

each country must exert a big effort to maintain a significant share in a small market, which tends to be increasingly competitive each year as transnational traders attempt to diversify supplies in tandem with other countries' attempts to penetrate them (Echeverri 1989).

A country such as Brazil—despite its reduced effective demand resulting from a highly skewed income distribution—can still use its own domestic market both as a springboard to exports and as a cushion to external shocks. But for a country like Guatemala, the world's largest exporter of cardamom but a small economy in itself, the recent glut in the market can be devastating (Brown and Suarez 1991). Of course, ceteris paribus, an expanding world economy should lead into an increased demand in any particular market, therefore reducing the intensity of competition among suppliers (Islam 1990).

Market size also can be considerably diminished by the protective barriers that usually form part of any country's economic policies. Until the erection of GATT, tariff barriers were the main hindrance to trade. Yet, after several GATT negotiations most tariff barriers—particularly in the processed and manufactured products—were substantially diminished as newer and increasingly sophisticated nontariff barriers were erected. Most fruits and vegetables, particularly when they are not processed, encounter less restrictive tariff barriers in the major industrial markets than do the mass-consumed "durable" food commodities (Lopez et al. 1989, Palmeter 1989, Garramon et al. 1990). By contrast, according to Nurul Islam; "Tariff rates escalate with the degree of processing" (1990:10).

In the United States, most fresh fruit and vegetable products enjoy duty-free treatment under the Generalized System of Preferences (GSP) or Most Favored Nation provisos of GATT. U.S. tariffs, however, still set stringent limits on Latin American exports of frozen concentrated orange juice, fresh cabbage, asparagus, broccoli, carrots, and cantaloupes (Knudsen and Nash 1990). Nonetheless, reversing its traditional position, in 1988 the U.S. government decided that several Latin American exports did not qualify for GSP treatment: dried apples from Argentina; guava, from Brazil; cut flowers from Colombia; and a number of fresh and preserved vegetables and some fruits from Mexico (Ballenger et al. 1991).

Besides tariffs, quotas, direct prohibitions, embargoes, and licensing—the visible restrictions—governments create many other barriers as part of their commercial policies. Plant and animal health restrictions are particularly problematic since it is difficult to establish clearly when a country is using these regulations as part of a legitimate concern for protecting its own production and population from pests and diseases and when these restrictions have become just another way to protect its farmers against potential competitors (Bright et al. 1987).

Since the early ninteteenth century there has been no avocado or papaya trade between the United States and the tropical countries bordering the Caribbean, because of the various pests claimed by U.S. authorities to be living in the region's groves. In 1984 an ethylene dibromine fumigant was approved then rapidly dismissed, since carcinogenic effects were discovered. In 1990, however, temperature controls and irradiation were approved, although these treatments reduce the quality and life of the product and decrease its competitiveness (Echeverri 1989, Smith 1992).

"Marketing orders" and the use of antidumping legislation are also mechanisms used increasingly for erecting nontariff barriers to imports. The 1937 U.S. Agricultural Marketing Act allowed farmer groups in any state to establish the internal quality standards and marketing regulations that would provide a marketing order for a particular commodity. Yet, a 1954 ammendment to the act, commonly known as Section 8(e), extended the same regulation to all imports. Marketing orders frequently are conceived as "farm programs that you don't see" (Powers 1990: 2).

Fourteen federal and twenty-nine state marketing orders regulated produce in 1987. Vegetable import regulations are currently in effect for fresh tomatoes, fresh potatoes, green peppers, cucumbers, eggplant, and fresh onions. As David R. Mares put it: "On the face of it, Section 8(e) appears to be a fair regulation, as it provides that both foreign and domestic producers must play by the same marketing rules. Nevertheless, certain production characteristics can be manipulated to discriminate against imports" (1987:205).

As is well documented in the literature, in 1968 the Florida Tomato Committee signed a marketing order setting up differentiated grades and sizes for mature green and vine-ripe tomatoes (Schmitz et al. 1981, Bredahl et al. 1987). Accordingly, "Florida producers grew some 90 percent of their tomatoes on the ground and harvested them green by machine. . . . In contrast, 90 percent of Mexican tomatoes were produced on stakes and hand-picked when ripe, a logical process given the country's comparative advantage in cheap labor" (Mares 1987:205). These provisos were promptly approved by the U.S. Department of Agriculture, establishing a de facto nontariff barrier on imports of Mexican tomatoes. As a consequence, in 1970 Mexican tomato production declined.

The first to react against these restrictions were the vegetable importers of Nogales, Arizona, who are exclusively supplied by Mexican producers. The West Mexico Vegetable Distributors of Arizona, a Nogales-based importer lobby with strong political support in the U.S. Congress, filed a restraining order to prevent the USDA from applying the marketing order. It was not until 1973, when the USDA bowed to legal and political pressure, that imported tomatoes were required only to comply to minimum grade and size and not to be graded and sized the same as Florida tomatoes (Bredahl et al. 1983:7). After long bilateral negotiations, the Mexican government agreed to restrict production and exports of tomatoes to the U.S. market, establishing a "voluntary export restrainment" (VER) in the international trade of nonprocessed agricultural products.

In 1978 various Florida grower associations, discontented with the Mexican VER, attempted to erect a new import barrier through the administrative procedures of U.S. antidumping legislation. Again, however, the Florida growers were unable to gain the necessary political support to succeed in their protectionist attempts. Therefore, despite the assessment of variable seasonally adjusted duties on fresh vegetables, the winter vegetable market between the two countries remains basically unhindered.

Other antidumping investigations have been more successful, however. Colombian and Costa Rican fresh flowers and ornamental plant exports have been submitted repeatedly to countervailing duties affecting their sales (Bulner-Thomas 1988, Ballenger 1991). In 1983 the U.S. International Trade Commission determined that the frozen concentrated orange juice imports from Brazil were "injuring," or "threatening to injure," its domestic industry. To offset the alleged subsidies, the government of Brazil agreed to impose an export tax on frozen concentrated orange juice exports to the United States (Ballenger et al. 1991).

On the other hand, Latin America is confronted with the unfair competitive practices of industrialized countries dumping agricultural commodities on the world market. The United States has four basic methods to increase exports: price reductions, provision of commercial credit, provision of food aid, and generic and branded export promotion programs. Credit and credit guarantees have been allocated to less-traditional U.S. exports, such as fruit juice and soft drink concentrates. By far though the most influential U.S. policies for protecting its fruit and vegetable sector are the export incentive and the market promotion programs.

The market promotion program (MPP) was authorized by the 1990 farm bill. It pledges $200 million annually from the USDA's Commodity Credit Corporation to help producers and traders finance promotional activities for U.S. agricultural products abroad. In 1991 nearly 50 percent of the MPP funds were destined for organizations promoting horticultural commodities. By contrast, only horticultural commodities have access to the export incentive program (EIP). Unlike the nonprofit participants in the MPP, the EIP finances private companies, such as frozen concentrated orange juice processors, table grape traders, and processed tomato exporters (Foreign Agricultural Service, August 1991).

Comparative Advantages: The Supply Side.[3] Natural endowments (climate, weather, soil fertility, water, and so on) as well as relative costs (salaries, transport, scale economies in processing and marketing, and learning processes) play a role in explaining the competitive edge that some countries may enjoy in the export of some commodities at particular junctures. They are the result, however, not only of immutable natural factors or unhindered market forces, but also of the policy dynamics among international and national entities. From an agro-ecological and climatic point of view, Costa Rica does not differ substantially from other Central American countries. Yet Costa Rica alone accounts for 33 percent of all horticultural trade in products other than fresh bananas and plantains entering the United States from the twenty-four CBI countries. To a large extent this pattern of trade is related to the country's long-standing reputation as a safe political haven for investment in the Central American region.

Several different elements enter into a coherent explanation of the emergence of export opportunities and their location in specific countries or regions.

Climate and agro-ecological conditions: Particular climate conditions, the quality and distribution of soils, and other agro-ecological characteristics—all directly related to a country's land fertility and potential product composition—have to enter into the explanation of the factors that directly affect investment decisions toward particular locations. For instance, Chile's climatic conditions in the Southern Hemi-

[3] Focus on "static" comparative advantages has been the strength of classical and neoclassical trade theories, old and new (Fransman 1986, Goldin 1990, Kuttner 1990). This theoretical perspective, however, has evident limits. Static comparative advantages (particularly those based on natural endowments) exist, as long as new technical innovations do not permit substantially lower costs in other locations. Therefore, mutatis mutandi, comparative advantages will go wherever technology goes.

sphere provided the opportunity to seize market niches in the Northern Hemisphere winter markets for fresh temperate fruit and vegetables. Different harvesting seasons in Brazil and the United States have also been the basis for U.S. orange juice processing and trading corporations' strategies to reduce storage costs and minimize risks through global sourcing (de Castro 1992).

A country's varied climate and soils, however, can be an asset for supplying its customers year-round as opposed to the seasonal supply of its competitors. Mexico's competitive edge over Israel and South Africa in the exportation of avocadoes to the European Community is based precisely on this climatic factor (Smith 1992).

Nonetheless, although land fertility, seasonality, and natural conditions might be important in a country's comparative advantages over potential competitors, the two most important cost components in the fruit and vegetable export industries are labor and transport costs.

Labor costs: Horticultural products are basically different from field crops in the capital and labor intensity of their production, processing, distribution, and marketing stages. High capital investments and intense labor requirements sometimes make the production of these crops unaffordable to potential producers, in spite of their natural comparative advantages. An inexpensive and abundant labor force and lower transportation costs, however, may compensate for high fixed capital investments.

One of the main comparative advantages of Mexico and the CBI countries in relation to their U.S. competitors is their access to cheap labor. Low labor costs can also partially explain the competitive edge of Chilean producers over other Southern Hemisphere suppliers who export to fresh winter markets in the Northern Hemisphere. In this latter case, however, the huge wage differentials are related not only to economic factors but also to the political demobilization of Chilean workers after the 1973 military coup. Whether Chilean fresh export industries will be able to afford the rising labor costs that will ensue from the increased unionization of farm laborers and the heavier taxation of social services in the current process of political democratization of the country remains to be seen (Gomez and Goldfrank 1991).

Transport costs: "Fresh" means perishable, as opposed to the durable characteristics of grains and other processed foods. This characteristic makes transport and insurance costs some of the biggest components in the cost structure of fresh fruits and vegetables. Not surprisingly, trade in horticultural products takes place mainly among neighboring countries or regions (Lopez et al. 1989).

Mexico and the CBI countries also benefit from their proximity to the huge U.S. market. Yet geographical proximity is only a potential benefit unless an efficient transport infrastructure is put into place. Transport costs are high for bulky items such as tropical roots (cassava, yams, sweet potatoes), converting proximity again into a competitive advantage. Mexico's success in penetrating the U.S. market for these products in relation to other potential suppliers in the Caribbean and South America is an example (Echeverri 1989).

Shifting competitiveness can also be the case if technical developments make transport costs less expensive. The increased competitive edge of Brazilian exports to the U.S. frozen concentrated orange juice market in the 1980s was partly a result of the development of tanker ships especially designed for this trade, instead of shipping the product in 55-gallon drums as was formerly the case (de Castro 1992).

Export infrastructure, which translated into on-time deliveries and better handling of the product, can be significantly related to the country's image of reliability among the international traders and the customers' acceptance of the product. By contrast, unreliable capacity allocation, unsuitable itineraries, inequality between imbound and outbound transport demand, and lack of port facilities constitute a hindrance (Lopez et al. 1989, Islam 1990). Chile's exports of berries, asparagus, cut flowers, and early table grapes to Europe are hindered not only by high airfreight charges but also by lack of shipping space (see Foreign Agricultural Service, May 1991). Similarly, Ecuador's fresh cut flower exports to the United States have been hampered by the lack of refrigerated rooms in which to store flowers awaiting transit in the airport. Interestingly, investments by a U.S.-based food trading corporation in strawberry production in Venezuela have been motivated by the firm's effort to lower its overall refrigerated ship cargo costs due to excessive outbound capacity in its trade with Venezuela.

Domestic market organization and international market skills: In spite of the free market rhetoric currently pervading most international trade literature, some kind of deliberate market organization seems to be a prerequisite to successful exports everywhere. In atomistic markets, in particular, economies of scale commonly exist that individual farmers are unable to realize. In those cases, some sort of coordination is still more pressing. Transnational corporations, controlling the distribution and marketing channels, provide one possible way of achieving coordination, yet there are other ways to do it.[4]

[4] In the United States, marketing orders were devised to help growers use collective action. Farm prices were historically low during the Great Depression of the 1930s. Several growers'

For example, the Mexican government, by stimulating the organization of Sinaloa's winter vegetable growers, "laid the foundation for avoiding early tendencies to respond to price increasing by oversupplying the market and thus depressing prices" (Mares 1987:24). By contrast, lack of organization has so far prevented Mexican avocado growers from successfully penetraing the U.S. market. Without a coherent organization. Mexican growers have to deal individually with oligopolistic packers and traders. As a result of the stringent conditions imposed by the latter, the growers prefer to sell their produce to domestic market traders, where market standards are less constraining, prices are more transparent, and risk is less (Smith 1992).

Finally, farmers must not only be aware of the opportunities offered by international markets but also must know how they work. Marketing skills are a key link between farmers and consumers. International fruit and vegetable markets are highly sophisticated. Lack of understanding of quality standards and governmental regulations, as well as the advantages of advertising and promotion abroad, may make the difference between losers and winners in these markets. To a large extent, Chile's success in nontraditional horticultural exports is related to a deliberate policy of adjusting supply to the diversified needs and tastes of current global markets, including a demand-driven market penetration approach based on scouting the world in search of export opportunities, as well as developing products well suited to the various domestic weather and natural conditions.

Striving for Competitiveness: State Intervention in an Export-Led Growth Strategy. Latin American governments have also been important actors in developing international competiveness in the emerging world order. Starting in 1973 in Chile, during the 1970s and 1980s most Latin American governments designed policies of open capital markets and flexible exchange and interest rates, while maintaining some selective import restrictions. The basic trend was toward export diversification, although this growth strategy did not dismantle anywhere the domestic productive activities, nor did it necessarily rule out all import-substitution policies. In most countries, therefore, a selective

co-ops attempted to raise prices by limiting sales to the market but were unsuccessful because nonparticipating farmers took advantage of the situation without bearing any of the costs of withholding produce from the market. In the spirit of the New Deal, the U.S. Congress passed the 1937 Agricultural Agreement Act to help farmers solve their "free-rider" problem. This act gave farmers' organizations considerable market power by making marketing order regulations for specific fruit, vegetable, and specialty crops compulsory to all handlers in a designated area (Powers 1990).

process of substituting imports emerged in tune with newly defined relative international prices. In agriculture, this mixed inward- and outward-looking strategy sometimes meant the reconquering of certain domestic markets, while promoting some nontraditional exports (Llambi 1990).

A turning point, however, was the Mexican moratorium on servicing its huge external debt in 1982. At that time, the international financial community—based on the principle of conditionality—provided the IMF with full authority for enforcing adjustment policies and for certifying eligibility for financial assistance on debtor countries (Gilpin 1987).

During the first half of the 1980s most Latin American governments, trapped between the external constraint of servicing the debt and the domestic constraint of maintaining a precarious social equilibrium, still had some authority to design a policy mix that alternated import substitution with outward-oriented growth. Therefore, under the umbrella of the "adjustment programs," austerity measures (to curb inflation and service the debt) and devaluations (to strengthen export-led growth and equilibrate the balance of payments) were coupled with public subsidies to become self-sufficient in agricultural products, strengthen the internal market, and protect domestic industries.

In the late 1980s, however, both national and international conditions deteriorated and a new turning point was reached, marked by the so-called Baker Plan. This time, the joint IMF/World Bank programs called for "shock" structural adjustment policies to remove subsidies and open the economy to market forces in the shortest possible time. As Robert Gilpin stated: "Debtors would take steps to open their economies to trade and foreign direct investment, reduce the role of the state in the economy through privatization, and adopt 'supply-side' market-oriented policies" (1987:326). It was implicit that in return the creditor nations would open their economies to debtor exports and increase debtor financing. As a result structural adjustment and not just short-term financial "adjustment and austerity" measures began to materialize. The structural adjustment package is based on three main policy prescriptions: (1) a competitive exchange rate through currency devaluations in pace with domestic and external inflation, (2) a commercial liberalization policy—with unilateral tariff reductions and elimination of all quantitative trade restrictions, and (3) financial liberalization to guarantee free entry of foreign direct investment in the most profitable areas of the economy.

The main rationale for undervaluing the national currency by means of regular devaluations is to promote exports and domestic savings. Yet, unless domestic inflation is curbed to parallel international rates, currency devaluations have the potential to increase the exporters' net returns—because it raises the prices they receive in domestic currency relative to their costs—only in the short run. In the longer run, however, imported inputs will also increase, eroding the initial advantages exporters achieved and thereby making new devaluations necessary to sustain competitiveness.

Besides, although devaluations may remove the anti-export bias of an overvalued exchange rate at least in some cases, the absence of state control over the exchange rate regime may negate the intent to stimulate nontraditional exports. This is the case of the oil-exporting countries of Latin America (Venezuela, Mexico, Ecuador, and Colombia), where, "if allowed to float according to market forces, petroleum revenues—which are produced at a very low cost in relation to the cost of other goods—will press the exchange rate toward overvaluation. The forces of the market tend to reestablish a rate of exchange which is overvalued, while stimulation of the non-oil economy depends on devaluating the currency to stimulate exports" (Beverly 1991:140).

On the other hand, the rationale for commercial and financial liberalization—the other two components of the structural adjustment program—is that (1) each country should specialize in the production of those export items that have evident static comparative advantages (natural resources, labor, and transport costs) in order to increase foreign exchange earnings and enhance import and debt payment capabilities; (2) foreign direct investment is necessary to finance the new technologies that will lead to efficient competition in world markets; (3) "if 'barriers' are removed then the 'correct' pattern of trade will ensue" (Kuttner 1990:39). All this is, of course, provided that the other part of the equation will someday materialize, that is, "that creditor nations would open their economies to debtor exports and increase debtor financing" (Gilpin 1987:326).

It is too early to evaluate the impact of the late 1980s–early 1990s structural adjustment policies in the Latin American economies, and particularly in the nontraditional horticultural export sectors, both because the process is too recent for all its effects to show and because there is still a paucity of empirical research on these issues. A future research agenda needs to include the impacts of structural adjustment and agricultural restructuring policies on both farmers and wage laborers, the impacts on nutritional standards both at the family farm level and at more aggregate regional and country levels, and the impacts of the new input-intensive crops on the environment.

It is clear, however, that devaluations, coupled with other adjustment policies such as labor, land, and capital deregulation are redefining each country's international competitiveness not only on the basis of agro-ecological or other natural criteria, but also on the basis of lower wage rates and land rents when compared with those prevalent in the industrialized world. During the seventies and eighties nontraditional fruit and vegetable exports grew in the context of sweeping agrarian counterreform, political demobilization of laborers, increased poverty, and financial takeover of the economy by transnational forces. Today, as it was yesterday, the impending question is still how far can these structural adjustment policies go without further explosive deterioration of the region's frail social and political equilibriums?

Future Prospects and Policy Options

Is nontraditional fruit and vegetable export-led growth dependable in the future? Or is this growth full of risks and uncertainties because it relies on highly volatile markets? Socioeconomic trends are always difficult to predict, as there are many uncertain and conflicting elements to take into consideration. This global transition, with higher market instability and increasing investment risks, makes prediction even more difficult. Therefore, rather than predict the future prospects of these markets, I will attempt to identify what the factors are that might affect their future performance, taking into consideration current trends and possible policy options.

Global Market Trends: Demand Side

The future growth of the aggregated fruit and vegetable export sector for each particular country will depend largely on global trends, such as the resumption of economic growth in the industrialized First World countries; the outcome of GATT negotiations; and the pace, direction, and transferability, of technological change. In considerations of particular markets and countries, however, contrasting and opposite tendencies have to be taken into account. In the current juncture, some international fruit and vegetable markets are facing increasing demand saturation, while others are growing at a rapid pace. At the country level, the basic instability of these markets means that commodity-growing regions rise and fall (Raynolds 1991).

In the United States—the largest and most proximate market for Latin American products—per capita consumption of fresh vegetables has been growing steadily, whereas the growth rate of processed vegetable consumption has been declining or stagnating (Cook 1990). Yet some winter fruit markets are showing signs of market saturation as extending growing and storage seasons in the Northern Hemisphere and new producing areas in the Southern Hemisphere are leading to increased competition and supplies (Gomez and Goldfrank 1991). By contrast, the potential market for most tropical fruits in the Northern Hemisphere's industrialized countries is still large, although to realize fully this potential will depend largely on the financial resources available to tropical exporters to set up more aggressive and innovative merchandising strategies. Finally, tapping market niches for specialty products—such as exotic spices, nuts, and flowers—is always a possibility, although these market segments reach their upper limits relatively rapidly.

Global Market Trends: Supply Side

Shifting competitiveness in the fruit and vegetable markets may also develop as a result of changing supply conditions in producing countries. In the 1980s land rent and water fees grew steadily in California and Florida as urban development put increasing competitive pressures on available space and natural resources. This trend will in turn tip the balance in favor of neighboring Latin American countries, where land rents and water fees are less expensive.

Shifting wage differentials as a result of policy decisions, themselves the result of social and political forces, may also enforce or diminish a country's comparative advantage in a particular market. In the United States, if the 1986 Immigration Reform and Control Act is not strictly enforced in the near future, the continuous supply of undocumented workers will further depress domestic farm wage rates, allowing U.S. horticultural growers to regain their competitive edge vis-á-vis Latin American competitors. By contrast, if labor market legislation is enforced, U.S. farmers will have no other option but to turn to enhanced labor-saving technologies or increased lobbying for protective legislation.

In the context of structural adjustment programs, devaluations made foreign direct investment in export-oriented horticultural activities more attractive as wage rates became increasingly inexpensive in dollar terms. But, depending on evolving economic and political conditions, shifting comparative advantages may also result from social pressures. In Chile, one of the consequences of the democratization process in the fruit export sector has been a relative loss of international competitiveness as temporary farm workers were given the right to strike farms and packing houses, in a context of decreasing U.S. labor rates (Gomez and Goldfrank 1991).

In most Latin countries, recent devaluations, subsidy cuts, and credit market liberalization policies, in a context of domestic inflation, have also resulted in increasing prices of both domestic and imported agricultural inputs and services and working capital. In Chile, high interest rates have cut into profit margins and have left many export-oriented growers in a risky financial position, endangering future growth potentials (Foreign Agricultural Service, June 1991).

Shifting relative input and product prices, as a result of both policy shifts and technical changes, may also be conducive to substitution effects between various fruit and vegetable products, as well as between a more and less traditional export mix. In Costa Rica, for instance, coffee—the country's largest agricultural commodity—competes with nontraditional exports for some agricultural inputs, such as labor. Thus growers are facing labor shortages for harvesting coffee, mainly as a result of a nontraditional horticultural export boom that is driving up wages (Arnade and Lee 1989).

New regulatory environments, both at the national and international level, also have the potential to significantly shift comparative advantages from one country or region to another. Safety concerns with chemicals and pesticide residues, for instance, may represent a major constraint to fruit and vegetable trade. These regulations in turn will lead to considerable advances in new environmentally safe technologies. Therefore, each country's ability to innovate or adapt alternative techniques of production will determine its ability to survive in the new market environment (Lopez et al. 1989).

Current trends in bloc formation may also affect shifting comparative advantages and competitiveness. It is still not clear, for instance, how the as yet undefined post-1992 "common" trade regime will affect imports from Third World countries other than the already preferential trade area linking some African, Caribbean, and Pacific (ACP) countries to the European Community. While the ACP countries are pushing strongly for continued preferential treatment, most Latin American countries are pressuring the EC to open its market. This struggle, which includes the interest of some U.S. transnational corporations with substantial investments in Latin America, is also shaping some current trade negotiations at GATT and beyond.

It is also not clear yet how the future North American Free Trade Agreement, between the United States, Mexico, and Canada, will affect U.S. trade relationships with the CBI countries. Nor is it clear whether the EC preferential trade programs for the Caribbean Basin and for Colombia, Peru, Bolivia, and Ecuador, eliminating tariffs and quantitative restrictions on exports to the community, will affect other processes such as President Bush's Initiative for the Americas, the Caribbean Basin Initiative, and the Andean Pact.

Finally, technical change, and particularly future developments in biotechology, which is currently at an experimental stage, will also affect competitiveness. In the long run, genetic engineering has the potential to increase substitution effects through the production of new products particularly suited to the sophisticated demand of affluent societies (Cook 1990). In the shorter run, however, the main risk for tropical Third World exporters coming from First World-led biotechnology research is the loss of potential markets due to the reduction of biological specificities of both crops and livestock products until now successfully grown only in the tropics. Bypassing the constraints imposed by climate- and location-specific products through genetically engineered new crops and livestock varieties may result in an increased profitability of First World producers, thus eliminating Third World natural comparative advantages.

Policy Issues: Adjusting Supply to Market Trends

Product diversification to already exported horticultural products may form part of an export-led growth strategy. Despite market saturation tendencies for some particular horticultural products, there are still large opportunities available in the development of new markets through carefully designed and properly targeted export promotion programs. The new trends in the international fruit and vegetable markets go in the direction of stressing high quality, producing fresh rather than processed goods, and developing agronomical practices that lead to products with no or reduced levels of pesticide and chemical residues (Lopez et al. 1989). Product diversification along these lines will require, however, not only finely tuned marketing strategies, but also consciously devised research and development programs at all stages of the production and marketing chains. This need is exemplified by Chile's strategy, discussed earlier, especially its development of a sophisticated network of domestic marketing infrastructure while establishing permanent trading offices abroad.

Another possibility is exporting value-added, industrially processed, agricultural exports. Canning, freezing, juicing, and manufacturing of jam and jelly are some of the opportunities still available to most already well-established fresh fruit exporters. But value-added agricultural exports are also capital- and knowledge-intensive industries; and these products also suffer frequent demand shifts and are protected by high tariffs and tariff escalation. Product diversification and value-added exports are not enough though. As signs of market saturation mount in some of the old tapped market outlets of the region, the development of new markets is needed.

The development of other markets besides the United States is a possibility. In a short-term perspective, only the increasingly wealthier economies of the Pacific Rim and Western Europe have the potential for becoming important consumers of Latin American fruit and vegetable exports. The wealthy countries of the Pacific Rim, in particular, are showing promising signs of becoming strong buyers for Southern Cone and Andean temperate fruit and Caribbean Basin tropical fruit exports, but—given the historical ties between Latin America and Europe further development of the Western European market is always a possibility. Yet these markets are extremely challenging, not only because of the distance and degree of sophistication that is required in order to supply them, but also because there is already strong competition with other Third World sources. In a longer run perspective, other markets (including Eastern Europe, the Middle East, and Southeast Asia) have potential, but their performance will depend largely on the resumption of global economic growth and the degree of openness of these regions' economies.[5]

Striving toward a More Balanced Growth Strategy

There seems to be no way back to the extremely biased import-substitution strategies with which most Latin American countries experimented in the past. It is clear that Latin America must search for a viable insertion in an emerging global order in which trade has become essential for technical advance in an increasingly integrated world. Yet

[5] Aware of its excessive reliance on the U.S. market, Chile's government and growers associations are trying actively to penetrate the European, Middle Eastern, and Pacific Rim markets. Yet given the locational advantages and historical links of powerful established exporters and the high protective barriers in these regions, these export markets have proved difficult to penetrate.

sustainable growth will not be achieved by following an equally biased outward-oriented growth strategy mainly based in the development of some well-priced—although highly volatile—horticultural export markets. For most Latin countries, such extremely unbalanced growth strategy would lead only to increased export vulnerability and import dependence.

A two-pronged approach adopting food security and selective industrial (i.e., agricultural and agro-industrial) policies is therefore needed, combined with coherently designed trade, monetary, and financial macroeconomic policies. Export promotion programs, in particular of nontraditional fruit and vegetable exports, may find a prominent place in this proposed policy scenario, although they do not make an end in themselves but should be developed having some broader policy goals in mind.

Even if export promotion programs were well designed and finely tuned with macroeconomic policies, they still may not be sufficient. In the face of trade barriers and bloc formation, would-be exporters may find it necessary to negotiate some market access for their nontraditional exports. More frequently than not, economically small and politically underprivileged countries, acting alone, are not in a good position to be successful in these bargaining activities with their giant trading partners in the industrialized world. Relatively small underdeveloped countries will have to make allies in the First World or develop workable alliances among themselves if they want to obtain significant market shares and afterward maintain their positions. As the Chilean and Mexican export experiences show, there is also a need to work in coalition with interest groups or lobbies from within the industrialized world to protect already achieved market shares (Mares 1987, Gomez and Goldfrank 1991).

These asymmetries in economic size and power are also reflected in GATT, in the relationships of the Third World countries with the leaders of the three main trading blocs currently developing, and in the relationships of debtor countries with the IMF/World Bank brokers of the transnational financial community. Enhanced bargaining power through political alliances in the First and Third Worlds also will be necessary to be successful in these other arenas of negotiation.

Conclusions

The shift toward outward-oriented growth strategies in Latin America has meant an effort to build up comparative advantages in off-season fresh fruits and vegetables, fresh cut flowers, and processed horticulturally based goods. These nontraditional exports are oriented mainly toward the high-income consumers and specialty market niches of affluent First World societies.

Although their markets are still not as large as those of the more traditional, mass-consumed, commodities that formerly composed the bulk of Third World exports, these new Third World exports are usually highly priced, making their market value outweigh their relatively reduced export volume. Furthermore, fresh perishable produce in particular confronts less restrictive tariff barriers than mass-consumed commodities. The dynamism of these commodity markets is based largely on the whims of consumer tastes in affluent societies for exotic commodities. But signs of saturation already are appearing, as First World competition is mounting, encouraged by nontariff barriers.

The development of nontraditional fruit and vegetable exports is not an easy task. Each activity demands not only the development of novel technical solutions but also highly sophisticated infrastructure facilities and marketing services. In the producer countries, domestic markets tend to be controlled by the oligopolistic power of a few transnational corporations, mainly through contract farming schemes. Sometimes this control is reinforced further by vertical integration into trading or marketing in the consumer countries.

So far, the bitter irony of the structural adjustment programs (imposed by the industrialized world through the IMF/World Bank/GATT on Latin American indebted nations since the 1970s) is that the search for comparative advantages in the agricultural realm has caused a striking waste of natural and labor resources, while poverty and hunger stalk the masses. The question, therefore, is how to include in the domestic self-sufficiency/international integration tradeoff the millions of hectares not cultivated as a consequence of restructuring, the waste in idle infrastructure, and the millions of farmers and workers currently unemployed or underemployed.

Nontraditional fruit and vegetable exports are by no means a panacea for the impressive problems facing Latin America in the current juncture. For most Latin American countries, to become specialized fruit and vegetable exporters may be nothing more than an updated version of former divisions of labor that reduced them to raw material exporters and manufactured goods and basic food importers. Are we heading back into the future?

References

Arnade, Carlos, and David Lee. 1989. *Risk Aversion through Nontraditional Export Promotion Programs in Central America*. Washington, D.C.: USDA–Economic Research Service.

Ballenger, Nichole, et al. 1991. *Agricultural Trade, Agreements, and Disputes in the Western Hemisphere*. Special report. Washington, D.C.: USDA.

Beverly, Tracy. 1991. "The Great Outward Turn." Thesis, University of Nebraska, Omaha.

Bonanno, Alessandro. 1991. "The Globalization of the Agricultural and Food Sector and Theories of the State." *International Journal of Sociology of Agriculture and Food* 1: 15–30.

Bredahl, Maury E., et al. 1983. "Technical Change, Protectionism, and Market Structure: The Case of International Trade in Fresh Winter Vegetables." *Bulletin* 249. Tucson: University of Arizona, Agricultural Experimental Station.

Bright, Brian, et al. 1987. *Trade Barriers and Other Factors Affecting Exports of Californian Specialty Crops*. Davis: University of California Agricultural Issues Center.

Brown, Richard, and N. R. Suarez. 1991. *U.S. Markets for Caribbean Fruits and Vegetables 1975–1987*. Washington, D.C.: USDA.

Bulner-Thomas, Victor. 1988. "El nuevo modelo de desarrollo de Costa Rica." *Revista Ciencias Economicas* 8(3): 51–66.

Cobban, Murray. 1988. "Tropical Products in the Uruguay Round Negotiations." *The World Economy* 11(2): 233–48.

Cook, Roberta L. 1990. "Challenges and Opportunities in the U.S. Fresh Produce Industry." *Journal of Food Distribution Research* February: 31–45.

de Castro, Monica. 1992. "The Dominican Orange Juice Industry: An Economic Analysis of Its Development and Prospects for the U.S. Market." Project report, Cornell University, Ithaca, N.Y.

Echeverri, Luis G. 1989. "United States Fresh Fruit and Vegetable Imports Industry: Industry Characteristics and Demand Analysis for Selected Tropical and Sub-Tropical Products." Project report. Cornell University, Ithaca, N.Y.

Foreign Agricultural Service. 1990–91. *Horticultural Products Review,* Various issues. Washington, D.C.: USDA.

Fransman, Martin. 1986. "International Competitiveness, Technical Change, and the State." *World Development* 14(12): 1375–96.

Friedland, William H. 1991. "The Transnationalization of Agricultural Production: Palimpsest of the Transnational State." *International Journal of Sociology of Agriculture and Food* 1: 48–58.

Friedmann, Harriet. 1991. "Changes in the International Division of Labor: Agrifood Complexes and Export Agriculture." In *Towards a New Political Economy of Agriculture,* ed. W. H. Friedland et al. Boulder, Colo.: Westview Press.

Friedmann, Harriet, and Philip McMichael. 1989. "Agriculture and the State System: The Rise and Decline of National Agricultures, 1870 to the Present." *Sociologia Ruralis* 19(2): 93–117.

Garramon, Carlos, et al. 1990. *La comercializacion de granos en la Argentina*. Buenos Aires: Legasa.

Gilpin, Robert. 1987. *The Political Economy of International Relations.* Princeton: Princeton University Press.

Goldin, Ian. 1990. "Comparative Advantage: Theory and Application to Developing Country Agriculture." Organization of Economic Cooperation and Development, Development Centre Technical Papers 16.

Gomez, Sergio, and W. L. Goldfrank. 1991. "World Market and Agrarian Transformation: The Case of Neo-Liberal Chile." *International Journal of Sociology of Agriculture and Food* 1: 143–50.

Islam, Nurul. 1990. *Horticultural Exports of Developing Countries: Past Performances, Future Prospects, and Policy Issues.* Research Report no. 80. Washington, D.C.: International Food Policy Research Instutute.

Knudsen, Odin, and John Nash. 1990. *Redefining Government's Role in Agriculture in the Nineties.* Washington, D.C.: World Bank.

Kuttner, Robert. 1990. "Managed Trade and Economic Sovereignty." *International Trade* 37(4): 39–51.

Llambi, Luis. 1990. "Transitions to and within Capitalism: Agrarian Transitions in Latin America." *Sociologia Ruralis* 30(2): 174–96.

Lopez, Rigoberto A., et al. 1989. "Constraints and Opportunities in Vegetable Trade." *Journal of Food Distribution Research:* 63–74.

Mares, David R. 1987. *Penetrating the International Market: Theoretical Considerations and a Mexican Case Study.* New York: Columbia University Press.

Palmeter, David. 1989. "Agriculture and Trade Regulations: Selected Issues in the Application of U.S. Antidumping and Countervailing Duty Laws." *Journal of World Trade* 23(1): 47–68.

Powers, Nicholas T. 1990. *Federal Marketing Orders for Fruits, Vegetables, Nuts, and Specialty Crops.* Agricultural Economic Report no. 629. Washington, D.C.: USDA.

Rama, Ruth, and Fernando Rello. 1979. "La internacionalizacion de la agricultura mexicana." *Estudios Rurales Latinoamericanos* 2(2): 199–223.

Raynolds, Laura. 1991. "The Restructuring of Export Agriculture in the Dominican Republic: Instability in Third World Commodities, Firms, and Labor." Paper presented at the 47th International Congress of Americanists, July 1991, New Orleans, Louisiana.

Sanderson, Stephen. 1986. *The Transformation of Mexican Agriculture: International Structure and the Politics of Rural Change.* Princeton: Princeton University Press.

Schmitz, Andrew, et al. 1981. "Export Dumping and Mexican Winter Vegetables." *American Journal of Agricultural Economics* 63(4): 645–54.

Smith, Amanda P. 1992. "The Battle for Green Gold: The North American Free Trade Agreement—The Implications for the Avocado Market." Project report, Master of Professional Studies, Cornell University, Ithaca, N.Y.

von Braun, Joachim, David Hotchkiss, and Maarten Immink. 1989. *Nontraditional Export Crops in Guatemala: Effects on Production.* Washington, D.C.: International Food Policy Research Institute.

White, Charmaine C. 1991. *The Political Effect of Stabilization and Structural Adjustment Programs on the Caribbean.* B.A. thesis, Amherst College, Amherst, Mass., pp. 1–58.

9

The Restructuring of Third World Agro-Exports: Changing Production Relations in the Dominican Republic

Laura T. Raynolds

Although market liberalization is the password for acceptance in the current period of economic restructuring, analysis of the concrete mechanisms of Third World agricultural restructuring points paradoxically to the fundamental role of the state in this process. Third World governments are not simply engaged in deregulating their national economies, in the sense of lifting the political restrictions on free market activity, but rather are involved in restructuring local economies to accommodate changing world economic and political conditions. For most Third World states, this process entails efforts to secure a place for local economic activity within changing international circuits of accumulation, inasmuch as the prospects for establishing a strong nationally oriented economy have, since the colonial era, been largely ruled out.

Given the historical primacy of export agriculture in much of the Third World, it is not surprising that changes in this sector have formed the locus of recent restructuring in many peripheral nations. Since the mid-1970s we have witnessed the widespread expansion of nontraditional agricultural exports and the breakdown of traditional agricultural export sectors, which have long integrated Third World countries in the colonial-based international division of labor. Yet this shift

I thank Philip McMichael, Gary Green, and Frederick Buttel for their helpful comments on this article and acknowledge the support of the Inter-American Foundation Graduate Fellowship Program and the Equipo de Investigación Social, Instituto Tecnológico de Santo Domingo, which made my field work possible.

should not be mistaken for a natural, or purely market-driven, evolution in export commodities. Rather, it is the result of an active political project of export substitution pursued by Third World states with the strong, and in some cases coercive, backing of international institutions. Nontraditional agricultural commodities—such as specialty horticultural crops, off-season fruits and vegetables, and ornamental plants—have been introduced in many peripheral nations to substitute for declining traditional agricultural exports and to take advantage of growing fresh food and luxury good markets in metropolitan centers.

This essay analyzes one such process of agricultural restructuring in the Dominican Republic, where fresh fruit and vegetable exports have been promoted to compensate for falling traditional agricultural export revenues. The Dominican state has been critical in delineating and encouraging the growth of these new exports, with the clear support of multilateral financial institutions, bilateral aid organizations, and regional trading partners. Yet the production of nontraditional agricultural commodities has proved highly unstable over the course of the 1980s. The roots of this instability appear on the one hand to lie in the investment, production, and marketing characteristics of the particular commodities and firms involved in this export sector. On the other hand, the inability of the Dominican state to guarantee the local conditions for accumulation in nontraditional agricultural exports clearly fosters this instability. In a world market characterized by high capital mobility, the failure of the Dominican state to maintain the social and economic conditions conducive to international accumulation has jeopardized the competitive standing of local production in international capital circuits. This essay suggests that the Dominican experience is not unique, but exemplifies the difficulties faced by peripheral states attempting to secure a stable position in an increasingly competitive and volatile world market on the basis of new agricultural exports.

The Restructuring of Latin American Export Agriculture

An important body of literature that has emerged since the 1970s tries to both document and explain the dramatic changes that have been taking place in the agricultural export economies of Latin America and the Caribbean. The theoretical debate has shifted along with changing conditions so that a historical review of this material provides empirical, as well as conceptual, insights. The focus of debate over the recent transformation of export agriculture in Latin America moved from a discussion of the new international division of labor to an analysis of processes of internationalization. The most insightful contemporary debate has shifted again, to focus on the national and international restructuring of agriculture.

Although the intellectual roots of our understanding of the new international division of labor lie in the work of Folker Frobel, Jurgen Heinrichs, and Otto Kreye on changes in global manufacturing, most of the central issues in this discussion were brought out in parallel research undertaken by Steven Sanderson on transformations in Latin American agriculture. Whereas Frobel, Heinrichs, and Kreye (1980) argued that the "classical" international division of labor, in which Third World countries were integrated into the world economy as suppliers of primary products, was being replaced by a "new" international division of labor based on the movement of low-skilled manufacturing to Third World locations, Sanderson (1985, 1986a, 1986b) asserted that in the new international division of labor Third World agricultural export sectors are transformed rather than eliminated. Improvements in transportation, preservation, and communication systems during the 1960s facilitated the geographical separation of agricultural production from major consumer markets and encouraged the growth of new agricultural exports. Latin America witnessed the rapid growth of beef, fruit, and vegetable exports to satisfy North American appetites and a boom in flower exports to adorn Northern dinner tables (Feder 1978, Burbach and Flynn 1980, Rama 1985). The increasing locational alternatives of production fueled the global sourcing activities of agricultural transnational corporations that, as Sanderson 1986b pointed out, could locate part or all of their production in areas with the lowest input costs. An essential aspect of this transformation involved the industrialization and rationalization of Latin American agricultural export production and its increasing integration into international agro-food chains (Sanderson 1986a:54).

Research on the new international division of labor highlights important changes in Latin American export agriculture, but this approach is overly narrow and structural in its focus. Rhys Jenkins (1984) and others have argued convincingly that internationalization is a wider and more contradictory process than suggested in most discussions of the international division of labor. This alternative perspective notes that the relocation of (agricultural and manufacturing) production to the Third World has been relatively modest and highly concentrated and that the internationalization of capital takes place through commodity and financial, as well as productive, circuits (Jenkins 1984:40).

Following this approach Miguel Teubal (1987) emphasized the ramifications of U.S. exports of subsidized agricultural inputs and commodities in increasing the use of modern agricultural technologies and undermining domestic food production in Latin America. David Barkin (1987) focused on how Mexican agricultural policies and production patterns adapt to, and internalize, international price signals. Understanding the ramifications of the deepening debt crisis is clearly central to an analysis of the internationalization of agriculture, since as Sanderson (1989) has pointed out, the international debt/foreign exchange crisis has become a critical factor shaping agricultural policies and exports throughout Latin America.

Though the internationalization of capital approach holds a great deal of promise in informing our understanding of changes in agricultural export production in Latin America, this approach tends to give insufficient attention to the state and to concrete processes of domestic restructuring. The autonomy of Latin American states typically is considered to be undermined by the spread of international commodity markets, the increasing power of transnational corporations and institutions, and the global impacts of policy decisions in the First World (e.g., Barkin 1987, Teubal 1987, Sanderson 1989). Although transnational forces clearly shape the activities, and even the nature, of Third World states, these states continue to play a fundamental and variable role in the transformation of export agriculture.

An insightful approach to integrating the activities of states into our analysis of ongoing international changes revolves around the concept of restructuring. Within a political economy framework, restructuring refers to the dynamic process of change in the relations between constituent parts of capitalist economy which is forged by the decisions and activities of concrete social actors (Kolko 1988:10, Lovering 1989:198). From this perspective, an analysis of how transnational financial institutions are reshaping agricultural export sectors needs to be balanced by an analysis of how particular Latin American states respond to, and attempt to reorient, these pressures given domestic political economic realities. Integrating an understanding of the state into our analysis of transformations in export agriculture requires recognizing the state's role in upholding the conditions for both domestic accumulation and international accumulation based on the combination of domestic labor, domestic and foreign capital, and domestic and foreign markets (Bryan 1987). As has been stressed by Alain Lipietz (1987) and others, the conditions for successful accumulation are not stable or given, but must be socially regulated. Clearly, the success of Latin

American states in securing the social and economic conditions conducive to domestic and international accumulation depends on their legitimacy as well as their ability to mediate conflicting class interests.

In addition to focusing attention on the complex roles of Latin American states in transforming export agriculture, this approach highlights the variable roles of other social actors in configuring the process of restructuring. Transnational institutions and both large transnational and small entrepreneurial firms have played key roles in the recent restructuring of Dominican agriculture. Though the actions of rural workers and peasant contract growers are considered, these social groups are found to be relatively powerless in shaping the restructuring of export agriculture at this juncture.

The Demise of the Traditional Agricultural Export Sector

The Dominican Republic exemplifies the colonial legacy of the Third World in the traditional international division of labor as a producer of low-value, undifferentiated, agricultural export commodities. The local economy has been dominated for centuries by the production of sugar and, to a lesser extent, coffee, cocoa, and tobacco for sale in, first, European and, then later, North American markets. The Dominican Republic has been a classic case of what is often referred to as a dependent agricultural export economy—where the national economy is dominated by a few primary products that are produced on the basis of extensive foreign investment and are destined for metropolitan markets—or more specifically in the Caribbean literature as a plantation economy, revolving around the production of sugar (e.g., de Janvry 1981, Mandle 1982).

During the 1980s the contribution of these traditional agricultural exports to the Dominican economy declined dramatically. As can be seen in Table 9.1 the value of sugar, coffee, cocoa, and tobacco exports fell from US$504 million to $310 million between 1979 and 1989—a decrease from 58 percent to 33 percent in the contribution of these four commodities to total export earnings.[1] This shift in the Dominican export economy involved a decrease in the value and volume of traditional agricultural exports as well as an increase in the relative contribution of other export sectors.

[1] Data on total exports exclude those from the free trade zones. This focus follows Dominican national statistical convention and reflects the very real separation between free trade zone activities and the domestic economy.

Table 9.1. F.o.b value of Dominican exports (million current US$)

	1979	1980	1981	1982	1983	1984	1985	1986	1987	1988	1989
Traditional	504	468	732	471	452	514	380	360	308	325	310
Sugar[a]	233	330	559	307	299	322	209	170	166	178	191
Coffee (crude)	143	52	63	89	76	95	94	113	63	64	60
Cocoa beans	74	51	44	52	55	70	58	60	66	64	44
Leaf tobacco	54	35	66	22	22	28	18	18	14	18	16
Nontraditional	92	111	116	91	80	118	122	171	166	176	195
Agricultural	18	22	24	21	23	34	29	45	35	33	34
Agro-industrial	40	48	47	36	32	41	44	58	52	63	76
Industrial	34	39	41	31	23	42	46	66	77	78	84
Misc.[b]	0	2	4	2	3	1	2	2	1	2	1
Minerals	272	381	336	195	250	239	234	192	241	408	422
TOTAL[c]	868	960	1184	757	782	871	736	723	715	909	927

Sources: CEDOPEX data reported in Boletin Estadístico (various editions 1980–89).
[a] Includes sugar's derivatives (e.g., molasses, refined sugar).
[b] Includes handicraft and reexport products.
[c] Excludes free trade zone exports.

The erosion of Dominican traditional agricultural exports must be understood within the context of shifting world agro-food markets. Three interrelated changes in agricultural commodity circuits have been particularly important in undermining the country's sugar export sector. The first involves an overall decline in the demand for sugar in First World countries as a result of increased health concerns and the large-scale substitution of sugar by high-fructose corn sweeteners and low-calorie alternatives (Goodman et al. 1987:133). The second change entails an increase in the protectionist policies adopted by First World countries in an effort to retain a substantial, and high-priced, market for domestic sugar producers (Harris 1987:128). A third and closely related change involves dramatic declines in world sugar prices in the mid-1980s, resulting from excess productive capacity and heightened international competition (Harris 1987:129).

The deterioration of the world sugar economy has had particularly negative repercussions for the Dominican Republic, given its high dependence on exports and its limited markets. The majority of Dominican sugar is exported to the United States under a strict quota system. Between 1983 and 1986, the U.S. quota for Dominican sugar dropped by 64 percent (from 492,800 to 176,710 short tons).[2] These quota reductions dramatically curtailed Dominican export earnings, since world sugar prices tend to be about 70 percent lower than U.S. quota prices (Harris 1987:139). As noted in Table 9.1 sugar export revenues declined from a high of US$559 million in 1981 to $191 million in 1989.

Though the market prospects for Dominican sugar may improve, it appears highly unlikely that the country will ever be able to resume its traditional reliance on sugar exports. The long-term revitalization of the world market for cane sugar depends on the growth of new import markets and the breakdown of Western European and North American protectionist policies that would increase imports of less expensive foreign sugar and slow down processes of substitutionism (e.g., Desmarchelier 1991). Yet, even if the world sugar market were to expand, the Dominican Republic would have trouble significantly increasing its market share since its competitive position as an international sugar

[2] As a result of declining U.S. production, the Dominican sugar quota was raised to 469,949 short tons for the period January 1, 1989–September 30, 1990 (equivalent to 268,542 tons for a twelve-month period). This quota was revised four times during this period, reflecting the vagaries of an import system based on estimated domestic production shortfalls (USDA-ERS 1990:57–58).

producer depends on the unsustainable exploitation of cheap Haitian field labor.[3]

The Dominican Republic's other traditional agricultural exports are of much less importance to the national economy than sugar, but they have experienced similar declines (see Table 9.1). International prices for coffee and cocoa have collapsed since the late 1970s in response to global oversupplies and weak agricultural commodity markets.[4] Despite rising export volumes, the earnings generated by Dominican coffee and cocoa exports dropped by 58 percent and 40 percent, respectively, between 1979 and 1989. The contraction in tobacco markets was even more dramatic, due partly to growing First World health concerns and shrinking demand for tobacco products. Between 1979 and 1989 the Dominican Republic experienced a 70 percent decline in the value of tobacco exports.[5] Though changing world agro-food markets appear to have played the strongest role in undermining these agricultural exports, customarily high Dominican export taxes further restricted production in this sector.

The Dominican state's recent export substitution project, which centers on the promotion of nontraditional agricultural and agro-industrial exports, must be understood in the context of this deterioration in the colonial-based agricultural export economy. It is in fact the collapse in the country's traditional exports that has led to the political definition and support of the nontraditional agricultural sector. Whereas the notion of a nontraditional agricultural sector is commonly used in government and development agency literature, the definition of which commodities are nontraditional is state-specific. In the Dominican Republic the nontraditional agricultural sector has been defined by state policies to include (1) all agricultural export commodities except coffee, cocoa, tobacco, and sugar and its derivatives and (2) all agro-industrial commodities, whether destined for domestic or foreign markets. This definition recognizes the increasing similarity between agro-industrial products (industrially processed agricultural goods) and all agricultural export commodities, which must be transformed (selected, packaged,

[3] Haitian migrant workers are highly exploited in the Dominican sugar sector. In addition to creating problems between the neighboring countries, this abuse jeopardizes Dominican access to U.S. markets under the Generalized System of Preferences, which outlaws imports from countries sanctioning the use of forced labor (Plant 1987).

[4] As has already happened with sugar, biotechnological advances in the food industry appear poised for the large-scale substitution of new industrial products for cocoa (Goodman, Sorj, and Wilkinson 1987).

[5] Tobacco export figures would be roughly 5 percent higher if the free trade zones were included (USDA-FAS 1988).

labeled, etc.) in accordance with international standards. Thus, for example, in the Dominican Republic fruits, such as bananas and oranges, which have been exported from the region for hundreds of years, are considered to be nontraditional, as are exports of root crops, such as yucca, which have been grown on the island since the precolonial era.

The Recent Rise of Nontraditional Agricultural Exports

Important global and domestic forces have shaped the growth of the Dominican nontraditional agricultural export sector. The international debt crisis increased the Dominican Republic's need for export revenues at the same time that rising protectionism restricted access to critical First World markets. These changes increased the influence of transnational financial institutions, bilateral donors, and major trading partners (particularly the United States) over national policymaking. Throughout the 1980s the Dominican state espoused an export-led development strategy consistent with the new neoliberal orthodoxy promoted by the international financial community. Yet the particular policies implemented in this area and their success in fueling nontraditional agricultural exports varied according to internal political economic conditions.[6]

Shifting U.S. policies toward the Caribbean in the 1980s played an important, though convoluted, role in promoting nontraditional agricultural exports from the Dominican Republic. Despite the facade of free trade rhetoric, these policy changes involved a reorganization rather than a weakening of U.S. restrictions on imports from the region. In the case of the Dominican Republic, the most significant change came from declining U.S. sugar quotas, which cost the country upward of US$200 million and heightened the economic necessity of expanding alternative exports (C/CAA 1988:9, USDA-ERS 1990:57–58).

The impact of the Caribbean Basin Initiative (CBI), in contrast, has been much less significant than has often been assumed, even though the initiative was purported to promote nontraditional exports.[7] The CBI encourages U.S. investments in the Caribbean Basin and grants duty-free access to the U.S. market to certain commodities. Yet the

[6] Though this discussion focuses on agricultural exports, Dominican export substitution policies also promote manufactured exports from the growing number of free trade zones (Investment Promotion Council 1987:3).

[7] The CBI was enacted initially under the Caribbean Basin Economic Recovery Act of 1983 (U.S. Public Law 98-67, Title II) for a twelve-year period; it was extended indefinitely in 1990.

items specifically excluded by the CBI, such as textiles, petroleum products, and sugar, are major regional products making up fully 23 percent of Caribbean exports (Pantojas Garcia 1988:23). This policy thus had only very limited success in strengthening regional trade circuits. Though the Dominican Republic appears to be the foremost CBI beneficiary, with exports to the United States under this program worth US$174 million in 1985, closer scrutiny finds that over 95 percent of these products were eligible for duty-free entry under previous trade legislation (Paus 1988:199). The CBI has facilitated the growth of Caribbean nontraditional exports, but its public relations impact clearly has been much greater than its economic effect.

The growth of nontraditional agricultural exports in the first half of the 1980s is most directly linked to intensive pressure on the Dominican state from international financial institutions concerned with the country's substantial foreign debt. As in much of Latin America, the deepening of the debt crisis in the Dominican Republic necessitated the reorientation of state policies according to the new monetarist/trade liberalization approach championed by the IMF and the World Bank (Wood 1989). This approach emphasizes reducing state spending and stimulating private sector investments, particularly in exports, as the way to increase economic growth and debt repayment.

A rigorous series of structural adjustment and stabilization policies were established under a 1984 IMF standby agreement, signed by the Dominican government as a precondition for debt rescheduling. The peso was devalued and exchange rates were decontrolled, effectively breaking down state control over the national monetary system. Meanwhile, public-sector spending on social and redistributive programs was slashed. These policies sent local prices soaring, kindling one of the most violent "IMF food riots" in the region, which claimed close to a hundred lives (Murphy 1987:250, Walton 1989:318).

The internationally sponsored structural adjustment program reduced state support for the production of basic foods for the local market and encouraged agricultural export production. Two important laws were established to subsidize private investments in nontraditional agricultural exports. The Export Promotion Law (Law no. 69) benefits nontraditional agricultural and industrial exporters, providing incentive payments and import duty exemptions on materials used in the production of export commodities (Investment Promotion Council 1987).[8] Between 1984 and 1986 goods valued at some US$223 million

[8] Law no. 69 defines nontraditional exports as all those except sugar, coffee, cocoa, tobacco, and minerals. Though the law was enacted in 1979, application requirements were so strict

were imported duty-free each year under this law (Unidad de Estudios Agropecuarios 1989:29). The law provided a particularly strong stimulus to nontraditional agriculture. In 1984, for example, more than 60 percent of exports benefiting from this law were of agricultural origin (Ceara Hatton et al. 1986:164–65). The second major policy incentive established was the Agro-industrial Law (Law no. 409), whose stated goal is to lead to the industrialization of traditional agriculture and the integration of agro-industrial complexes (Banco del Comercio 1982:28). This law grants income tax and import duty exonerations from 40 to 100 percent to firms involved in the production of agricultural commodities that are standardized, packaged, or processed—in other words all nontraditional agricultural exports, as well as agro-industrial products for domestic and foreign markets.[9] According to the official registry, eighty-six agribusinesses benefited from Law no. 409 from its initiation in 1983 through 1986.

In addition to major tax exonerations, programs were initiated in the mid-1980s to provide state-subsidized credit to nontraditional agricultural exporters. These credit programs operate through the development arm of the Central Bank, the Fondo de Inversiones para el Desarrollo Economico, but were established and funded largely by international institutions. The Inter-American Development Bank and the U.S. Agency for International Development provided roughly US$107 million in loans to establish these programs, with smaller counterpart funding coming from the Dominican government (JACC 1990:15).

The most direct support the Dominican state provides to the nontraditional agricultural sector comes from the inexpensive rental of state land that is being retired from sugarcane production. Due to the collapse of the sugar economy, two of the state-owned sugar mills were closed in the mid-1980s, and roughly 13,000 hectares of state land have been taken out of sugarcane cultivation. About 10,000 hectares of this land have been made available to agribusinesses through the Sugar Diversification Project, which receives the majority of its funding from the U.S. Agency for International Development (USAID 1987).

The export promotion policies initiated by the Dominican state, in conjunction with transnational financial institutions and bilateral do-

that only three firms qualified for exonerations in 1980; when the restrictions were eased in 1984 as part of the export promotion push, the number of firms receiving exonerations grew to 233 annually between 1984 and 1986 (Unidad de Estudios Agropecuarios 1989:29).

[9] The exact level of exoneration depends on the location of the enterprise, the share of raw materials used that is of national origin, and whether the firm is vertically integrated.

nors, greatly stimulated the growth of the nontraditional agricultural sector in the first half of the 1980s. Foreign and domestic investment in this sector grew substantially, as did the value and volume of exports. From 1979 to 1986 the export earnings of Dominican nontraditional agricultural and agro-industrial exports rose from US$58 million to $103 million (see Table 9.1). The greatest increase during this period came from exports of fresh agricultural commodities, whose value rose from US$18 million to $45 million. Major exports in this area included (1) tropical root crops, particularly taro, yams, and yucca, (2) vegetables and horticultural crops, particularly peppers, tomatoes, green beans, squash, and eggplants; and (3) tropical fruits, particularly melons, pineapples, and avocadoes (CEDOPEX 1987). The value of frozen or processed agro-industrial exports increased more slowly during this period, but continued to account for more than half of the total export earnings of the nontraditional agricultural sector. Beef products were the most important and fastest growing agro-industrial exports, followed by canned and frozen beans, cocoa products, and processed coconut (CEDOPEX 1987).

Despite early successes in state-sponsored export promotion, the nontraditional agricultural export sector stagnated during the second half of the 1980s. In fact, the value of fresh agricultural exports dropped significantly, from a high of US$45 million in 1986 to an annual average of $34 million in subsequent years (see Table 9.1). Though this decline is partially accounted for by falling prices for some important exports, the volume of nontraditional agricultural exports in most cases fell even further (CEDOPEX 1988, 1989). As shown in Figure 9.1, the export earnings of tropical root crops and vegetables and horticultural crops declined precipitiously. The contribution of these commodity groups to the total value of nontraditional agricultural exports dropped from 32 to 20 percent, and 37 to 15 percent, respectively, between 1986 and 1989 (CEDOPEX 1988, 1989). Exports of some tropical fruits also fell, but substantial increases in pineapple exports offset these losses, and as a whole the earnings of this commodity area continued to rise. Exports of many agro-industrial goods also dropped in the late 1980s, though this pattern was less pronounced.

One of the major reasons for the decline in nontraditional agricultural exports appears to be that the Dominican state was no longer able to guarantee the conditions for accumulation in this sector as a result of mounting debt and state fiscal crises. A leading Dominican economic commentator has characterized the period between 1986 and 1990 as one of economic chaos and deinstitutionalization, where for domestic

Figure 9.1. F.o.b. value of Dominican nontraditional agricultural exports

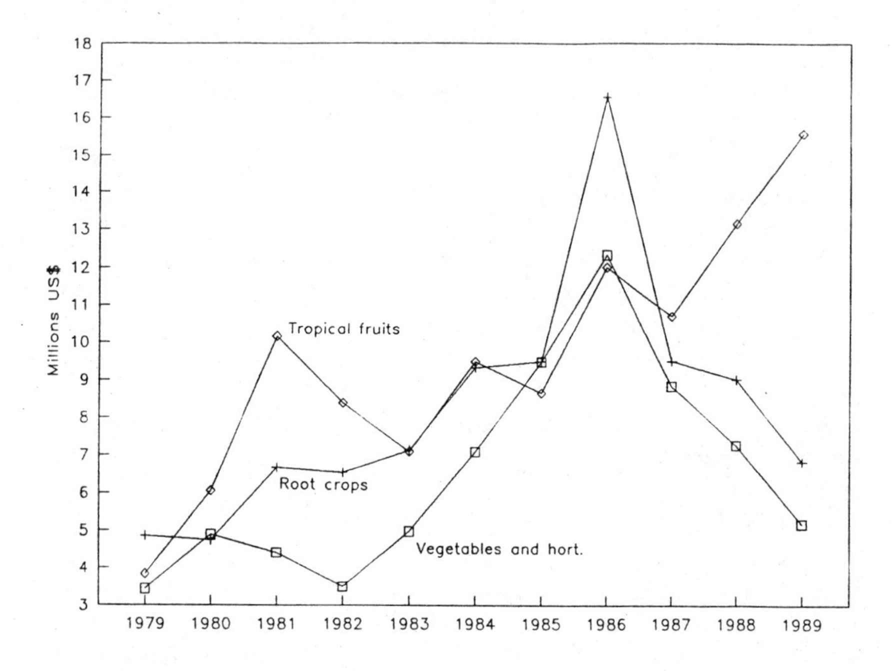

Source: CEDOPEX data reported in Boletin Estadístico (various editions 1980–89).

political reasons the government focused the limited public funds available on urban-based construction (Ceara Hatton 1990). During this period the state suffered severe fiscal difficulties fueled by the unavailability of new international loans and attempts to pay off at least a portion of the outstanding national debt. Though the Dominican government refused to sign a formal IMF agreement in the latter half of the 1980s for fear that the urban riots of prior years would be repeated, informal negotiations with the IMF were maintained (U.S. Dept. of Commerce 1989).

As the Dominican state continued to voice its support for the export-led development strategy advocated by the international financial community, it cut back its expenditures and regulatory activities in nontraditional agriculture (as well as almost all other areas of the economy). Export promotion policies were not legally revoked, but the majority of incentives were not being honored by the late 1980s and were at best selectively applied. In 1989 the government ceased paying important incentives for nontraditional agricultural exporters which had been provided under the Export Promotion Law (Unidad de Estudios Agropecuarios 1989:20–22). Nontraditional agricultural exporters felt that they were so unlikely to receive legally stipulated import duty exonerations under the Agro-industrial Law that many no longer bothered to register for them.[10]

In sum, although nontraditional agricultural commodities came partially to replace traditional Dominican agricultural exports during the 1980s, the growth of this sector was erratic. The Dominican state implemented export promotion policies in accordance with the priorities of the international financial community, but state subsidies to domestic and foreign investors were not fully maintained because of the fiscal crisis and changing domestic political priorities. The state's inability to guarantee the conditions for accumulation in the nontraditional agricultural sector undermined a prediction by *The Economist* that between 1986 and 1991 the Dominican Republic would be "the Caribbean country which offers by far the most important agricultural and agro-industrial [investment] opportunities" (Burgaud 1986:35).

The Production of Nontraditional Agricultural Exports

The ramifications of changing political-economic conditions in the production of nontraditional commodities in the Dominican Republic can be better understood by turning from a discussion of national trends to an analysis of particular agricultural export commodity systems. The key area of growth in the nontraditional agricultural export sector in the 1980s was in fresh produce—particularly tropical fruits, vegetables and horticultural crops, and tropical root crops. The introduction and expansion of these nontraditional commodities entailed changes in the relations under which agricultural commodities are both produced and marketed (Raynolds 1993).

In the Dominican Republic, fresh fruits and vegetables destined for export are not usually purchased on the open market, because of their perishability and the need for extensive coordination in their production, shipping, and marketing. Instead, two alternative production systems predominate: either growers are contracted to produce com-

[10] The erosion of official support for Law no. 409 was brought to my attention by agro-industry managers and was substantiated by my finding that, of the 123 firms previously registered to receive benefits under this law, only 35 bothered to renew their applications in the first half of 1989.

modities on their own land (an out-grower model) or workers are hired to cultivate land controlled by their employer (a plantation model). A brief review of these two forms of production and of the specific experience of crops grown under these conditions in the Dominican Republic helps demonstrate the important role of the state in maintaining these agricultural commodity systems.

The out-grower model is characterized by a large number of small firms that limit their fixed investments and maximize their production flexibility by contracting with peasant growers to produce nontraditional commodities. This contracting system is commonly utilized in the Dominican Republic in the production of agronomically sensitive and labor-intensive horticultural crops, since the exporting firm is thus able to shift a large portion of the high production risks and costs onto peasant producers. In this model, agribusinesses typically provide (1) technological assistance; (2) modern seeds, pesticides, herbicides, and fertilizers; (3) production credit; (4) internationally acceptable processing, packaging, and shipping; and (5) access to international marketing chains. Actual production is carried out by peasant growers using predominantly household land and unpaid family labor.

The experience of oriental vegetable production in the Dominican Republic demonstrates the operation and inherent instability in this contract system. Oriental vegetables—including such crops as Chinese eggplants, snow peas, fuzzy squash, hot peppers, and yenchoy—were first introduced by entrepreneurial capitalists of Asian origin, who by the early 1980s had established over a dozen small export enterprises. These export firms were able to maintain relatively low levels of investment (averaging between US$160,000 and $800,000) by limiting their expenditures on land and labor. Although a few firms produced some of their own vegetables, most relied on production contracts with a total of 2,000 to 3,000 peasant producers (*Listin Diario*, August 14, 1989, sec. D, p. 1). Through these contracts, exporting firms were able to control their produce supply and take advantage of ecologically diverse agricultural regions, while limiting their exposure to the high production risks and heavy labor requirement of horticultural cultivation. Roughly half of all production risks and costs were thus passed on to local peasant households.[11] Oriental vegetables were shipped to brokers

[11] The distribution of costs depends on how local land and labor resources are valued vis-à-vis access to specialized inputs, knowledge, and markets. Given the labor-intensity of oriental vegetable production, it is fairly clear that the contributions of growers are at least equal to the contributions of exporting firms. Since product prices are set at the time of shipping, the risks inherent in market price swings are shared equally.

in the major eastern ports of the United States to be sold through minority marketing connections to growing Asian migrant communities and ethnic restaurants.

At its height in 1986 the export of oriental vegetables accounted for roughly 13 percent of total nontraditional agricultural export earnings of the Dominican Republic;[12] however, by 1989 exports of many of these commodities had virtually disappeared. The majority of oriental vegetables produced in the Dominican Republic have been denied entry into the United States, their major market, because they have persistently failed to meet standardized import requirements. Between 1987 and 1988 the U.S. Food and Drug Administration placed Dominican shipments of major oriental vegetables under automatic detention, because of findings of excessive pesticide residues (Murray and Hoppin 1992). In 1989 several more oriental vegetables were restricted entry to the United States, this time because of pest infestations (*Listin Diario,* August 14, 1989, sec. D, p. 1).

The crisis in the oriental vegetable production system highlights two major foci of instability in the nontraditional agricultural export sector.[13] First, small exporters often do not fully understand, nor can they easily meet, international import restrictions. Although U.S. sanitary and phytosanitary regulations embody real health and safety concerns, they also act as important nontariff barriers to horticultural trade (Islam 1990). Second, exporting firms concerned with short-term profits may promote intensive production practices that generate pest and pesticide problems, since they do not bear the long-term costs of these actions (Murray and Hoppin 1992).

Between 1988 and 1989 major oriental vegetable exporters reported a 57 percent decline in cultivated area, with estimated losses of roughly US$16 million (JACC 1989:27–28). Yet the Dominican state did little to manage the crisis. Despite its declared commitment to supporting the conditions for private capital accumulation in nontraditional exports, the Dominican government dedicated few resources to resolving these

[12] The accurate calculation of oriental vegetable exports is hampered both by ubiquitous underreporting as a means of avoiding taxation and by the grouping of oriental vegetables with like products in national statistics. My estimate is based on official CEDOPEX data and includes unambiguous commodities and half the value of grouped products known to be made up substantially of oriental varieties (i.e., peppers, eggplants, green beans, and cucumbers).

[13] Although the technical characteristics of oriental vegetables (including their perishability, short growing cycle, and susceptibility to pests) clearly contributed to the difficulties, what is of interest here are the social origins of the crisis.

production problems.[14] The failure of the state to monitor and control production conditions within its borders jeopardized both the Dominican Republic's position as a major supplier of oriental vegetables and domestic food security, since food crops are being threatened by pest problems originating in this export sector.

The state was slow to respond to this crisis, but oriental vegetable firms acted quickly, demonstrating the production flexibility and geographical mobility inherent in contract production. Since most firms had limited fixed investments, they were able to close down their operations virtually overnight as profits disappeared. Enterprises run by first-generation Asian migrants with few local ties left the Dominican Republic within a year to start again in neighboring countries having no oriental vegetable export restrictions.[15] Firms with greater local ties focused on diversifying into unregulated markets and new commodity areas. The lack of state control over the geographical ability of oriental vegetable firms resulted in a loss of contracts for many Dominican peasant households, which were left with few, if any, profits and no employment or viable markets for their products.

The second major model used in the production of nontraditional agricultural export commodities in the Dominican Republic follows more conventional plantation lines. Most large transnational corporations operate under this system, hiring workers to cultivate land managed by the company, thereby assuring complete control over the production and supply of produce. Though these enterprises bear similarities to the plantations of centuries past, current processes of economic and political restructuring have fundamentally reshaped this pattern of production. A central feature of the new plantations being established in the Dominican Republic is that they use state land that is being taken out of sugarcane cultivation.[16] Under the internationally sponsored Sugar Diversification Project, transnational corporations have been able to gain access to state land, thereby limiting their fixed investments and increasing their production flexibility.

One of the most important and rapidly growing areas in which this form of production has been established in the Dominican Republic is

[14] The U.S. Agency for International Development spearheaded the initial efforts to identify pest and pesticide problems at the point of export.

[15] Though oriental vegetables shipped from the Dominican Republic were restricted, exports from another country, even if exported by the same firm, entered the normal system of sporadic customs inspections.

[16] After the 1961 fall of Trujillo, the Dominican state appropriated the dictator's vast sugar plantations, thereby becoming the owner of twelve of the country's sixteen sugar mills and their associated lands.

the production of pineapples. In 1980 pineapple exports contributed only 1 percent of nontraditional agricultural export earnings; by 1989 they made the largest contribution to fresh produce exports, with 14 percent of the total value (CEDOPEX 1983, 1989). Two transnational corporations, Dole Food Company and Chiquita Brands (previously United Fruit Company), account for the lion's share of these exports. Both companies produce on state land, and interviews with managers confirm that the availability of state lands was one of the most important reasons for locating their operations in the Dominican Republic. Chiquita has leased roughly 1,000 hectares of state sugar land at a concessionary rent on the basis of a long-term renewable contract. Dole has access to some 2,000 hectares of state land through a joint venture agreement that accords the Dominican government a percentage of the total profits from the enterprise. The problem with this type of joint venture is that transnational subsidiaries calculate profits using rather arbitrary internal transfer pricing conventions, and there is no reason that a particular subsidiary need ever show a book profit.

In the context of global production options, the growth of pineapple exports from the Dominican Republic is attributable to the availability of cheap labor as well as state land. Dominican wages are among the lowest in the Caribbean region (Bobbin Consulting Group 1988). In order to encourage foreign investment, the state has kept wages low by setting minimum wage rate increases below the rate of inflation and creating a special "subminimum" wage for agribusinesses and companies in the free trade zones. Most of the roughly 2,000 workers in the new pineapple plantations are casual day laborers who are unprotected by national labor legislation. These workers, many of whom are women, have no job security and are paid less than even the subminimum wage. Labor unions in this sector have been either crushed outright or co-opted by the combined forces of the state and the transnational corporations.

Although these new plantations have higher fixed costs than enterprises using contract production systems, pineapple corporations maintain a great deal of flexibility through their global sourcing operations. By maintaining subsidiaries in diverse locations—including Hawaii, Thailand, the Philippines, Guatemala, and Honduras—Chiquita and Dole can guarantee access to the lowest cost produce, as well as compensate for natural crop fluctuations. These corporations are able to play one production site against another, thereby increasing their bargaining position when negotiating concessions with individual governments. In the Dominican context, promises of cheap land and labor have attracted pineapple corporations by reducing the required fixed investments of subsidiary firms. Yet these very concessions increase the likelihood that production will relocate if the state does not maintain a satisfactory level of subsidization. Though the conditions for accumulation in this sector appear to have been sustained thus far, the Dominican state has forfeited direct control over critical national land resources and rural labor forces.

The experiences of oriental vegetable and pineapple production systems help illustrate the tenuous nature of the growth of the nontraditional agricultural export sector in the Dominican Republic. State-subsidized access to land, export incentives and tax exonerations, and guaranteed access to cheap labor clearly have been important in the development of this sector. Yet the collapse of oriental vegetable exports illustrates the very real limits to the Dominican state's ability, both administratively and financially, to maintain the conditions necessary for private capital accumulation. The state appears at present to be upholding the conditions for accumulation in pineapple production, but these conditions are repeatedly challenged by competing demands on state resources.[17] The nature of the firms involved in these nontraditional agricultural exports feeds the instability of this sector, since despite their seemingly diverse organizational forms they are all organized so as to minimize their fixed costs and maximize their production flexibility and their mobility (Raynolds 1993). The instability in the commodities and firms involved in the nontraditional agricultural export sector leads to a fundamental instability in the labor forces that are integrated into this new sector.

Conclusions

This essay demonstrates the critical role of the state in the restructuring of Dominican export agriculture. The country has experienced a very important shift from a dependence on colonial-based agricultural exports to a more diversified export economy in which nontraditional agricultural and agro-industrial commodities play a major role. The growth of these new exports has taken place under the political direction, and active encouragement, of the state. With strong backing from multilateral financial institutions, bilateral aid organizations, and ma-

[17] For example, in 1990 government commitments to lease public land to agribusinesses were annulled in order to satisfy peasant claims for agrarian reform land and thus garner votes in a close presidential race.

jor trading partners, the Dominican state delineated and subsidized nontraditional agricultural exports as part of its export substitution strategy. The importance of the state's role in supporting this sector, and maintaining the conditions for international accumulation based on local agricultural production, is most strikingly evident in analysis of the instability experienced in nontraditional exports in the 1980s.

The roots of this instability are located partially in the investment, production, and marketing characteristics of the particular commodities and firms involved in the nontraditional agricultural sector. International fresh fruit and vegetable markets are particularly volatile because of the perishability, sensitivity, and seasonal nature of these commodities. Yet it is clear that the organization of the firms in this sector reinforces this commodity-based volatility. Global sourcing options establish high levels of competition between alternative production sites. Firms can take advantage of this competition, either, in the case of small firms with limited fixed investment, by remaining footloose and relocating their operations when necessary or, in the case of transnational corporations, by shifting production between subsidiaries and using this mobility as a negotiating point in gaining concessions from particular governments. The flexible character of the firms involved in the Dominican nontraditional agricultural export sector is not specific to this context, but reflects the growing importance of production flexibility in the survival of capitalist enterprises operating in increasingly globalized markets.

Ultimately, it is Third World states that are responsible for mediating this instability and securing a place for local economic activity within changing international circuits of accumulation. States must not only establish the local social and economic conditions conducive to accumulation, but must also maintain those conditions, since firms have the geographical mobility to take advantage of better productive opportunities elsewhere. In the Dominican case the 1980s debt/fiscal crisis undermined the capacity of the state to guarantee, and socially regulate, the conditions for accumulation in nontraditional agriculture, prompting a decline in investments and export earnings. As demonstrated in the oriental vegetable sector, what is at stake here is not only short-term foreign exchange earnings and domestic employment, but the long-term productivity of critical natural resources.

The predicament for Third World states attempting to secure the conditions for domestic and international accumulation in the current conjuncture is greatly exacerbated by international pressures that argue for deregulation yet, paradoxically, require national regulation. The insistence by multilateral banks and bilateral aid organizations that Third World economies be deregulated and their national boundaries relaxed fuels the production mobility of firms. Yet, as demonstrated in the Dominican Republic, under conditions of locational firm mobility states are required to provide private sector subsidies to attract investments and actively must manage local productive resources so as to ensure the long-term viability of local production in international accumulation circuits.

Even as debt-ridden Third World states increasingly are obliged by the international financial community to "open" their economies and promote exports, even the World Bank (1986:24) concedes that rising First World tariff and nontariff barriers to trade sabotage export-oriented development strategies. As exemplified in the Dominican case, rising protectionism in metropolitan markets has been a central factor undermining traditional agricultural exports and necessitating the promotion of nontraditional commodities. Yet new agricultural exports appear to face equally tough and politically volatile barriers to trade.

Although this chapter draws on the experiences of the Dominican Republic to illuminate the important dimensions of state action in the current restructuring of export agriculture, these conditions do not appear unique. Conjunctural forces, including the increasing competitiveness of international markets, the heightened mobility of firms, and the international pressures for market deregulation, create difficulties for all peripheral states attempting to enter rapidly changing world agrofood markets. Exactly how particular states negotiate this transition, and their success in securing a place for local economic activity in international agricultural circuits, may vary. Yet, regardless of these potential variations, Third World states clearly are implicated in the ongoing process of restructuring.

References

Banco del Comercio Dominicano. 1982. *Ley 409 sobre incentivo y protección agroindustrial.* Santo Domingo: Dr. Milton Messina y Asociados.

Barkin, David. 1987. "The End of Food Self-Sufficiency in Mexico." *Latin American Perspectives* 14(3): 271–97.

Bobbin Consulting Group. 1988. *Sourcing: Caribbean Option.* Columbia, S.C.: Bobbin Consulting Group.

Bryan, Richard. 1987. "The State and Internationalisation of Capital: An Approach to Analysis." *Journal of Contemporary Asia* 17(3): 253–75.

Burbach, Roger, and Patricia Flynn. 1980. *Agribusiness in the Americas.* New York: Monthly Review Press.

Burgaud, Jean-Marie. 1986. *The New Caribbean Deal*. Special Report no. 240. London: *The Economist* Intelligence Unit.

Caribbean/Central American Action (C/CAA). 1988. "In Caribbean Basin, Sugar Is a Tough Habit to Break." *Caribbean Action* 1: 9.

Ceara Hatton, Miguel. 1990. "De Guzman a Balaguer: de la 'demanda inducida' a la 'reactivacion desordenada.'" *El Siglo* (Santo Domingo) March 31: sec. D, p. 6.

Ceara Hatton, Miguel, et al. 1986. *Hacia una reestructuración dirigida de la economía dominicana*. Santo Domingo: Portada de Taller.

Centro Dominicano de Promoción de Exportaciones (CEDOPEX). 1979–89. *Boletin estadístico*. Santo Domingo: CEDOPEX. Various editions.

——. 1979–89. Departmento de Estadísticos, preliminary data for 1989. Santo Domingo: CEDOPEX.

de Janvry, Alain. 1981. *The Agrarian Question and Reformism in Latin America*. Baltimore: John Hopkins University Press.

Desmarchelier, John. 1991. "Changing Structure and Cycles in the World Sugar Industry." *Sugar y Azucar* March: 18–26.

Feder, Ernest. 1978. *Strawberry Imperialism: An Enquiry into the Mechanisms of Dependency in Mexican Agriculture*. Mexico City: Editorial Campesina.

Frobel, Folker, Jurgen Heinrichs, and Otto Kreye. 1980. *The New International Division of Labor*. Cambridge: Cambridge University Press.

Goodman, David, Bernardo Sorj, and John Wilkinson. 1987. *From Farming to Biotechnology: A Theory of Agro-Industrial Development*. New York: Basil Blackwell.

Harris, Simon. 1987. "Current Issues in the World Sugar Economy." *Food Policy* May: 127–45.

Investment Promotion Council of the Dominican Republic. 1987. *Investing in the Dominican Republic*. Santo Domingo: Investment Promotion Council.

Islam, Nurul. 1990. *Horticultural Exports of Developing Countries: Past Performance, Future Prospects, and Policy Issues*. Research report no. 80. Washington, D.C.: International Food Policy Research Institute.

Jenkins, Rhys. 1984. "Divisions over the International Division of Labor." *Capital and Class* 22: 28–57.

La Junta Agroempresarial de Consultoría y Coinversión, Inc. (JACC). 1989. "La producción y exportación de vegetales chinos." *Agroempresa* November-December: 26–27.

——. 1990. *Agroempresa Dominicana*. Santo Domingo: JACC.

Kolko, Joyce. 1988. *Restructuring the World Economy*. New York: Pantheon Books.

Lipietz, Alain. 1987. *Mirages and Miracles: The Crisis of Global Fordism*. Trans. David Macey. London: Verso.

Lovering, John. 1989. "The Restructuring Debate." In *New Models in Geography: The Political-Economy Perspective*, ed. Richard Peet and Nigel Thrift, pp. 198–223. London: Unwin Hyman.

Mandle, Jay. 1982. *Patterns of Caribbean Development*. New York: Gordon and Breach Science Publishers.

Murphy, Martin. 1987. "The International Monetary Fund and Contemporary Crisis in the Dominican Republic." In *Crises in the Caribbean Basin*, ed. R. Tardanico, pp. 241–59. Newbury Park, Calif.: Sage Publications.

Murray, Douglas, and Polly Hoppin. 1992. "Recurring Contradictions in Agrarian Development: Pesticide Problems in Caribbean Basin Nontraditional Agriculture." *World Development* 20(4): 597–608.

Pantojas Garcia, Emilio. 1988. "The Caribbean Basin Initiative and the Economic Restructuring of the Region." In *The Other Side of U.S. Policy towards the Caribbean: Recolonization and Militarization,* pp. 23–33. Río Piedras: Caribbean Project for Justice and Peace.

Paus, Eva. 1988. "A Critical Look at Nontraditional Export Demand: The Caribbean Basin Initiative." In *Struggle against Dependence: Nontraditional Export Growth in Central America and the Caribbean,* ed. Eva Paus, pp. 193–214. Boulder, Colo.: Westview Press.

Plant, Roger. 1987. *Sugar and Modern Slavery: A Tale of Two Countries.* London: Zed Books.

Rama, Ruth. 1985. "Some Effects of the Internationalization of Agriculture on the Mexican Agricultural Crisis." In *The Americas in the New International Division of Labor,* ed. Steven Sanderson, pp. 69–94. New York: Holmes and Meier.

Raynolds, Laura T. 1993. "Institutionalizing Flexibility: A Comparative Analysis of Fordist and Post-Fordist Models of Third World Agro-export Production." In *Commodity Chains and Global Capitalism,* ed. Gary Gereffi and Miguel Korzeniewicz. Westport, Conn.: Greenwood Press.

Sanderson, Steven. 1985. "The 'New' Internationalization of Agriculture in the Americas." In *The Americas in the New International Division of Labor,* ed. Steven Sanderson, pp. 46–68. New York: Holmes and Meier.

——. 1986a. "The Emergence of the 'World Steer': Internationalization and Foreign Domination in Latin American Cattle Production." In *Food, the State, and International Political Economy,* ed. F. LaMond Tullis and W. Ladd Hollist, pp. 123–48. Lincoln: University of Nebraska Press.

——. 1986b. *The Transformation of Mexican Agriculture: International Structure and the Politics of Rural Change.* Princeton: Princeton University Press.

——. 1989. "Mexican Agricultural Policy in the Shadow of the U.S. Farm Crisis." In *The International Farm Crisis,* ed. David Goodman and Michael Redclift, pp. 205–33. London: Macmillan.

Teubal, Miguel. 1987. "Internationalization of Capital and Agro-industrial Complexes: Their Impacts on Latin American Agriculture." *Latin American Perspectives* 14(3): 271–97.

Unidad de Estudios Agropecuarios, Consejo Nacional de Agricultura. 1989. *Lineamientos para una estrategia de desarrollo selectivo de exportaciones prioritarias.* Santo Domingo: Unidad de Estudios Agropecuarios, Consejo Nacional de Agricultura.

U.S. Agency for International Development (USAID). 1987. *Sugar Diversification Project File.* Santo Domingo: USAID.

U.S. Department of Agriculture, Economic Research Service (USDA-ERS). 1990. *Sugar and Sweeteners* March: 57–58.

U.S. Department of Agriculture, Foreign Agricultural Service (USDA-FAS). 1988. *Dominican Republic—1988 Annual Agricultural Situation Report.* Santo Domingo: USDA-FAS.

U.S. Department of Commerce. 1989. *Omnibus Trade Act Report on the Domini-*

can Republic—Final Text. Washington, D.C.: U.S. Department of Commerce. Telegram.

Walton, John. 1989. "Debt, Protest, and the State in Latin America." In *Power and Popular Protest: Latin American Social Movements,* ed. Susan Eckstein, pp. 299–328. Berkeley: University of California Press.

Wood, Robert E. 1989. "The International Monetary Fund and the World Bank in a Changing World Economy." In *Instability and Change in the World Economy,* ed. A. MacEwan and W. Tabb, pp. 298–315. New York: Monthly Review Press.

World Bank. 1986. *World Development Report 1986.* Washington, D.C.: World Bank.

PART IV

COMMON, CONTRADICTORY, AND CONTINGENT FORCES

10

Is the Technical Model of Agriculture Changing Radically?

Pascal Byé and Maria Fonte

The conjunction in the 1970s of the economic crisis and major breakthroughs in the field of genetic engineering led many authors to think that agriculture was on the verge of a technical and scientific biorevolution (OTA 1986, Goodman et al. 1987, Joly and Zuskovitch 1990). The large-scale diffusion of the new technical model would reverse the trend toward decreasing productivity and increasing production costs, so it was thought. Biorevolution would push toward the homogenization of agricultural production conditions to industrial forms, causing a sharp drop in prices and opening new markets for agricultural products in a context of price deregulation.

These analyses were based on different foundations. Some authors stressed the importance of the scientific and technical paradigm shift, from a mechanical-chemical to a biological model (OTA 1986, Goodman et al. 1987, Bonny and Daucé 1989); others pointed to the relevant role of deregulation policies (CESTA 1987). Overall technological forecasts were very optimistic, envisaging the overcoming of any technical and economical limit to the existing model of production and the solution of any food, energy, and environment problem. They, however, rested implicitly upon one relevant explanatory variable to the exclusion of the others and their interrelations.

Since the latter 1980s, among the social scientists in the United States, the debate on biorevolution has focused on the evolutionary or revolu-

We thank an anonymous reviewer and Philip McMichael for their comments, which have greatly helped in improving the clarity of the exposition. This research was partially supported by funds from the Italian MURST—fondi 60 percent.

tionary nature of change (Buttel 1989, Otero 1991b). We emphasize that this debate is not so much about the speed of change, as about the scope of applicability and impact of the new technology and the qualitative nature of change.

In Frederick H. Buttel's opinion we are witnessing the evolution of a technology whose applicability and impact are limited to declining sectors. More specifically, in agriculture "biotechnologies are not likely to differ greatly from previous technical forms on the basis of their 'revolutionary' productivity impacts" (Buttel 1989:12), though they will differ as far as intersectoral reorganization is concerned, since they will bring more vertical integration and homogenization. For Gerardo Otero, instead "to the extent that new technologies are at the core of the world economic restructuring, biotechnology will play a major role in transforming agriculture. . . . Like the Copernican revolution, biotechnology will lead to unforeseen developments" (1991b:560–61). Another possibility is that biotechnology "might provide the world with an option to steer the future of agriculture in a sustainable-regenerative direction, while maintaining and increasing the productivity achievements of modern agriculture" (Otero 1991a:1). Though everyone recognizes the potentiality of biotechnology in this direction, critical social scientists believe this path of biotechnological development to be very unlikely in today's social and economic context.

It is useful to state the problem of the "qualitative nature of change" in the following, more explicit, terms: authors are tying to determine whether biotechnology will lead to an evolutionary growth of the agro-food system, as we know it, with an increasing subordination of agriculture to industry and its organization in product-specific *filiéres* (commodity systems), or to a radical change in the economic, social, and technical interrelations of the agro-food system, leading to its full industrialization.

In a position that is partially self-critical, Buttel (1989) points out that, twenty years after the discovery of recombinant DNA, the diffusion of biotechnology in the agro-food system is very slow, its impact is restricted to specific areas and sectors, and commercialized products are still very few. Social scientists in general (Fanfani et al. 1992, Junne 1992) tend to revise the interpretation of the "revolutionary potential of biotechnology" in favor of a more realistic analysis of its short-term impact. More attention is given to the obstacles and limits to the diffusion of biotechnologies.

This debate can be a useful starting point for a discussion of the current agricultural transition, focusing on the technical dimension.[1] Its evolution needs to be located in the context of the decline and crisis of Fordist model at the end of the twentieth century. Up to the 1970s new technologies, such as biotechnologies and information technologies, tended to be incorporated into the existing industrial-intensive model of production without upsetting it. Innovations are selected according to the interests of industrial organizations that dominate the commodities systems, but as soon as the crisis of the Fordist structures of production reaches agriculture (in the 1980s—about a decade after affecting industry [June 1992]), the trajectory of biotechnology development becomes less clear and more indeterminate.[2]

In light of the above-mentioned debate and in an attempt to give account of the limits to biotechnological development, our aim in this essay is to contextualize biotechnological evolution since the early 1970s. Our position is characterized by various assumptions.

First, in contrast to those who see a linear derivation of techniques from science—deriving the idea of a necessary biorevolution from the discovery of recombinant DNA—we believe that technical change is an interactive process, with continuous feedbacks between its upstream and downstream phases (research, development, production, marketing). In the eighties a change in the characterization of food demand and in the social functions of agriculture in developed countries may be seen, for example, as the basis for a change in the trajectory of biotechnological evolution.

Second, innovation processes and technical change may be better interpreted by a systemic approach, in the context of evolutionary theory.[3] There is no radical change up to the point at which any eventual limit or constraint to the evolution of the system can be overcome through adjustments within the current techniques. So, for example, technical discontinuities in the agro-food complex stimulate the concentration and delocalization of production (Bonnieux and Rainelli 1988); increases in land prices, especially in the vicinity of big cities, work in favor

[1] Incidentally, we can observe an analogy with the discussion of the "agrarian question" at the end of the past century (Kautsky 1959).

[2] The length of the diffusion process is not only the expression of the adjustment process between demand and supply forces, but also of social conflicts, contradictions, and resistances among different actors, according to their unequal capacity to appropriate or utilize new technical acquisitions.

[3] We accept here the distinction between atomistic-mechanistic models, which view systems as consisting of parts and relations between parts, neither of which changes over time, and evolutionary thinking, which looks at systems as undergoing changes in their parts and relations (Norgaard 1992). Since the 1970s the evolutionary theory has successfully been applied to innovation theory in economics (see Dosi et al. 1988).

of intensification; supply control policies exert pressure toward an increasing artificialization of crop and animal production. The adjustments set in motion by the constraints are local and circumscribed. They are not transmitted in a cumulative way to the whole system and do not concern its whole nature. It is only the adding or rearrangement of functions or both that may change radically the technical system.

Of the many different obstacles to the development of biotechnology, we give emphasis in this essay to the increasing heterogeneity in the agro-food system of Western postindustrial countries, deriving from technical and socioeconomic maladjustments and discontinuities.[4] These refer to contradictions between old Fordist mechanical-oriented techniques and new post-Fordist techniques based on biological and biochemical innovations, as well as to contradictions between the traditional functions of agriculture in economic growth and its new emerging functions in the postindustrial society. We identify this movement as a trend toward a rehabilitation of diversity and flexibility in agro-food production.

Biotechnology may be a useful technical tool in mastering the heterogeneity of constraints and their differential impacts on agro-food complexes. A major role in this process will be played by the chemical sector, because of its flexibility, its multisectoral and transnational organization, and especially the strategic position of its corporate actors, who are able to integrate into the same knowledge base the efficiency of scale economy and the advantages of differentiations deriving from economy of scope. They may, then, be able to dominate the contradictory movements between heterogeneity and homogenization of techniques, between Fordist norms and the transition toward more environmentally friendly agro-food policies, evolving toward flexible forms of industrial production.

This process will not lead in the foreseeable future to a biorevolution in agriculture, but rather to more acute contradictions and conflicts between the persistence of a technical model of production still oriented toward quantitative norms and the emergence of new demands and new functions for agriculture and the rural space.

A Trend toward Heterogeneity in Agro-Food Production

Although other authors in this volume (Friedland, Stanley) document the trend toward an increasing homogenization of the production proc-

[4] We recognize however the working of an opposite trend toward homogenization that is well documented in other parts of this volume (see Chapters 5 and 7).

ess in the agro-food system, we stress a parallel movement in postindustrial societies of the Western world toward increasing heterogenity. Maladjustments between techniques derived from both mechanical and biological forms of knowledge and new agricultural policies, addressing more general, macroeconomic objectives, lead to a greater variety of adjustment strategies, which in turn will increase the heterogeneity of adopted technical solutions.

Capital Immobilization and Technical Discontinuities

Adoption on a large scale of industrial techniques in agriculture is based on product standardization on one side and homogenization and segmentation of increasingly specialized production processes on the other (Friedmann and McMichael 1989). This evolution has been accompanied by the substitution for fragmented and polyvalent forms of labor and land (Weber and Maresca 1986, Sigaut 1989) of homogeneous and specialized forms of capital—land, machines, intermediate goods, skills (Benvenuti 1990, Van der Ploeg 1990). Mechanical techniques largely have dominated this process of substitution, both directly through the substitution of machines for the work force and indirectly through the subordination of other agricultural techniques (e.g., fertilization, breeding, plant protection) to mechanical norms. Plant breeding, for example, has selected progressively for mechanical features in cultivars (e.g., hard tomatoes); cultural practices have eliminated complex forms of rotations and mixed farming; chemical fertilization has displaced organic fertilization and complementarities among cereals and leguminouses. Obstacles to mechanization in animal agriculture have not impeded the mechanization and industrialization of feedstuff preparation and other operations in intensive livestock farming.

The supremacy of norms deriving from a mechanical logic, which has led to extreme forms of farm concentration and specialization, has nonetheless left unresolved the major impasse of the heavy capital immobilization. Investments in machines in fact substitute for live labor, increase labor productivity, and release workers from toil, but fail to change radically the empirical nature of a production process, which remains essentially a complex biological and natural process (Mann and Dickinson 1978), technically based on a long-standing accumulation of experience (Byé and Fonte 1992). Grain production, for example, still requires a set of tools that reproduces the empirical form of work and individually substitutes for hand operations: soil preparation, plant care, harvesting. The same is true for forage production, where machines are multiplied by the complexity of harvesting, storage, and processing operations. The amortization of each specialized machine is more difficult to realize with polyculture, where specialized implements must be multiplied in correspondence with the mechanization of any animal or crop production operation. Mechanical techniques poorly master the complexity of living forms and processes.

In order to minimize risks and reduce costs, generally in a context of price support policies, farmers tend to specialize production and increase the size of the farm. But if choices for rationalization and economies of scale may solve temporarily some of the basic problems, they induce land concentration and capital immobilization (Servolin 1972) on the one hand and irreversibility of investment choices on the other. Mechanical techniques, then, have a cumulative effect in terms of fixed capital and financial investments, which tend to increase disproportionately, and have considerable negative effects in terms of supply rigidity in relation to market and price evolution. At the very moment in which economic and social constraints work in favor of adaptive capacities and flexibilities, large specialized farms can not in fact easily switch from one crop to the other in a short time. The problem is not specific to agriculture, it also concerns upstream input industries and downstream transformation and distribution activities. Food industries need to differentiate their production in order to satisfy an increasingly volatile and segmented demand, as economic deregulation and stricter health and environmental regulations compel the input industries constantly to adapt their products.

In more general terms, the technical discontinuities weaken the robustness of the agro-food chain. Heterogeneity of techniques—in and between both agriculture and agro-food industries—remains a serious constraint in agro-food production all over the world, even in the most industrialized agro-food sectors. The problem has long been attenuated by the adaptive role played both by farmers and their empirical knowledge in relation to the use of most rigid tools or machines (Byé and Fonte 1991), and by the extension service with respect to the application of scientific knowledge to empirical practices. In the face of the disappearance of farmers' adaptive knowledge and practices, the permanence of technical heterogeneities and discontinuities becomes a more serious problem. Two points of discontinuity are particularly relevant.

In agriculture, the most intensive forms of production (e.g., greenhouses cultivation, soilless agriculture, irrigated farm systems, confinement housing, and feedlots) tend to question the coherence of technical combinations built around mechanical norms. Technical and economic difficulties in the process of mechanization both of specific crops and animal husbandry (vegetables, fruit, greenhouse crops, or animal production) and of specific operations (plant protection, harvesting specialty crops, pruning) testify to the inadequacy of mechanical monolithism. They also highlight technical discontinuities, which persist as long as some practices, crops, and production systems are mechanized and others are not. These discontinuities contribute to calendar bottlenecks as well as market disruption, where norms and grading standards are concerned.

Such inconveniences to capital and profit accumulation might be reduced through the substitution of either biological (new seeds, new varieties) or chemical objects for mechanical objects, in specific points at first, then progressively to all parts of the production process. But this substitution will inevitably lead to the decline of mechanical norms or their progressive adaptation to biological processes. Competition between the various subsectors of the agro-input industries—machines and fertilizers, phytosanitary and seed industries—emerges inevitably also in the substitution process. New practices of pest management may in fact imply a change in the traditional practices of crop cultivation or soil preparation, substituting biochemical inputs for mechanical operations.

In the food-industry, maladjustments between generic techniques applied to staple food (feed, sugar, and so on) and the persistence of specific empirical techniques applied to specialized filières (dairy, quality food) are growing. Techniques applied to processes of transformation, preservation, conditioning, packaging, and distribution are still heterogeneous, revealing their craft origins. They are defined in relation to specific products: dairy, meat, beverages. These techniques are still pervasive in the world and are an obstacle to the diffusion of modern techniques (fractioning, separation, recombination), which tend to standardization and which are the basis for the expansion of the commodity systems and the new industrial models of food production (Byé, Frey, and Monateri 1990).

The Reorientation of Price Systems and the Emergence of New Social Functions for Agro-Food Systems

Farm support systems of the postwar era in metropolitan states are unraveling. Deregulation, erosion of the peasant constituency power, liberalism, monetarism, agricultural and farmland surpluses, and concern for a better quality of food and a better management of natural resources are finally transforming agricultural price support systems. These are on the way to losing their specificity and increasingly integrate new values as the expression of a more general political economy: monetary and interest rate policies, international trade, exchange rate and income policies, subsidies and international cooperation (Joslin 1985), and space and environment management (Commission des Communautés Europèennes 1988).

Specific agricultural constraints deriving from the peculiarities of the agricultural sector are progressively disappearing, as other generic external constraints are emerging, testifying to the progressive integration of agriculture into global trends. Enlarged references for technical patterns, for example, must take into account the progressive liberalization of national markets, the use of interest rates in anti-inflationary policies, the instability of exchange rates, the transformation of natural resources (up to now freely available) into public goods whose costs of conservation and reproduction (Green and Yoxen 1990) are not taken into account by market mechanisms, and the emergence of new social needs in the sphere of leisure, health, and security (Byé and Fonte 1991).

Variations that follow such a general reformulation of agricultural prices affect agricultural investments in various forms. They may have only a weak influence on investments initiating at the time of accelerated growth—which, following mechanical norms, tend to increase capacity and, notwithstanding the crisis, are retained—but a strong influence on current investments, facing uncertainties in financial and international markets and policies.

Modifications of prices systems imply then the reopening of the range of technical choices. Valorization and maintenance of farmland surpluses (set-asides, management of natural spaces, reconversion of intensive agriculture) cannot in fact be guaranteed by the simple transposition of techniques that were originally aimed at increasing quantities. A reduction in the use of the traditional inputs (fertilizers, phytosanitary products), for example, cannot be a long-term solution for pollution problems; minimum tillage and zero tillage, though alleviating soil erosion, may not be adequate for the management of set-aside land; extensive stockfarming does not consist simply in avoiding crowding or reducing high-density confinements.

The adoption in the EC of quota policies aimed at limiting the supply of items such as milk, sugar, and wine is first and foremost an incentive to rationalization investments (control systems, automatic devices, and the like) that can extend the lifespan of existing equipment and techniques. But these quota policies are not an adequate answer to the requirements of new production norms. Generalization of quota policies will then induce in the long run the restructuring of livestock farming and the reformulation of the demand for genetic, feedstuff, and other technical objects.

Extension of premium prices for quality imposed by the food industries under the pressures of a qualitative change in demand for food (Fanfani, Green, and Rodriguez-Zuñiga 1992, Wilkinson 1992) and new health and environment protective norms contradicts previous orientations in techniques toward low-cost mass production. The adjustments involve an extensification of industrial techniques (minimum tillage, controlled use of fertilizers, integrated pest management) and the increasing control of residues in farm and agricultural products. The continuous pressure of quality norms implies the transformation of the status of the agricultural products. Rather than mass commodities, they will become health and medical goods, in the context of a market more strongly regulated by quality, health, and environmental norms. Such a paradigmatic transformation could be the basis for a new model of production.

In sum, although Fordist approaches to agricultural production (increasing volumes and standardization of products) tend to reduce technical choices to standards, multiplying discontinuities in the technical model of production and raising market and currency uncertainty, emerging new functions for agriculture, new agricultural policies (aiming at the control of supply and taking into account problems of management of renewable resources), and the growing attention to the demand for quality food (both from a health and organoleptic point of view) tend to increase the variability and instability of technical adjustments. New possibilities are open for incremental innovations as well as for technical breakthroughs.

Toward Industrial Flexibility: Interpreting the Biotechnological Evolution

Recent evolutions in the agro-food system lead agriculture toward a multipurpose model of development, expressing itself in heterogenous forms of production, techniques, and knowledge bases. The capacity to master this heterogeneity, by making production forms more flexible, is the real stake for industrial actors upstream and downstream of the agro-food commodity system. Biotechnology and information technologies are useful tools in this process.

Differentiation and Flexibility of Technical Choices

Despite the ongoing process of homogenization in agricultural techniques, the above-mentioned evolutions in the agro-food system tend to change the prevailing function of agriculture—the production of standardized goods in increasing quantities and at decreasing costs for the agro-food industry and the urban markets—adding diversified, complex objectives. These objectives may appear to contradict Fordist logic: management and reproduction of natural resources (in particular farmland surpluses created by the concentration of production in increasingly restricted parts of the territory), preservation of rural communities, and protection of health, quality, and the environment.

Differentiations in the technical model tend to respond to these changes. Differences are accentuating between extensive techniques and capital-intensive techniques. The former are land-using–labor-saving techniques, which are not only persisting but also are spreading because of a marked drop in land prices, the reduction of the work force, and agricultural policies such as the implementation of set-aside policies or environment government regulation. The latter integrate the new price constraints and promote a better utilization of natural resources, such as water, light, and soil quality (Ruttan 1990). In opposition to the preceding model, we can qualify these models as "land-saving–water-and-light-using."

These two models coexist in the EC, where extensive techniques predominate in the cereal grain basins and capital-intensive techniques concern mainly the specialized crops in the south and west. They coexist also in the United States, where the expansion of extensive farms in the different "belts" does not hinder the strengthening of the intensive agriculture based on more efficient techniques of water and space management in the southern and western regions.

These models, both pursuing the optimum performance from land, do not have identical conditions of transformation. The *land-using* model is strictly dependent on government policies (set-aside, management of natural reserves) linked to regulated price systems and founded on extensive Fordist mechanical-chemical techniques. The diffusion of this model depends on a general fall of land prices or easy credit conditions in bouyant markets. Conversely, the implementation of new price systems favors diffusion of the *land-saving* model. In a situation of decreasing agricultural prices, this model, technically based on biochemical norms and principles, guarantees higher productivity increases. Arising in conditions of demand instability and market segmentation, it uses new forms of organization, production, and marketing, responding to new agro-food system configurations and new social demand for specialized foods.

When the conditions of development of the two models are compared, it is clear that they evolve according to different norms and goals and that they occupy different positions in the innovation system. The first, based in a price system dependent largely on politically constructed national farm sectors, utilizes the standardized Fordist technologies of the postwar era. The second benefits from innovations and technology transfers from noncommercial sectors (health care, defense and nuclear research, telecommunication), emerging in a context where farm supports are undergoing deregulation and food demand is increasingly differentiated.

But as long as both models of production are based on the exploitation of nonrenewable resources (i.e., hydrocarbons)—that is, on the same foundation and paradigm that led to the artificialization of agro-food techniques—they do not solve the emerging problems of ecosystem management and health, quality, and environment protection, perpetuating the contradictions between the co-evolution of technical and ecological systems (Norgaard 1992). Diversification of techniques then increases their flexibility, but the principles of efficiency, which underlie them, are not compatible with sustainability.

Are Technical Choices Irreversible?

As the uncertainty of market and productivity trends and the substitution of capital for labor increase, the agricultural demand for industrial inputs changes progressively. Long-term investments (land, machinery, buildings, and capital goods) decrease, as short-term expenses for industrial inputs (fertilizers, animal feeds, pesticides, and other chemical products) increase. This substitution process is governed by the *irreversibility* of new technical choices, which amplifies the consequences of technical change, and by the *inertia* effects of past choices, which, on the contrary, restrain them.

Inertia effects are determined largely by the technical and social choices realized during previous periods. To a great extent, they govern the rhythm of technical change in agriculture, insofar as they encourage avoidance of all innovations that would accelerate the technical and social obsolescence of existing goods and know-how and integrate all the innovations that can improve the efficiency of existing formal or informal investments. The success of herbicides or treatment products in the agricultural mechanical model, for example, depends primarily on the fact that before posing the problem of an alternative to the current Fordist model they complement it.

Irreversibility effects are revealed in the inelasticity of agricultural demand for certain industrial products. They favor the utilization of goods and services linked by a common technical factor. This inelasticity of demand is particularly noticeable when producers use new genetic varieties and new breeds. Starting from a new animal breed or a new seed, farmers are involved in a technical process that is determined by the very characteristics either of the breed (nutrition, growth regulation, health, marketing conditions) or the seed (fertilization, weed control practices, pest management, harvest, transformation, marketing).

"Inertia" and "irreversibility" do not refer to two distinct and mutually exclusive sets of innovations. A radical change may be introduced, first, in order to stabilize a conventional technique or to compensate a handicap derived from the rigidity of a previous technique. It can even be deployed to overcome an interruption or a limitation in the diffusion of a given current technique, that is, in the context of an inertial kind of strategy. Indeed the radical innovation can be aimed initially at the development of a product or the diffusion of techniques already dominant in the market. The creation of new seeds resistant to herbicides, for example, may be aimed initially at the opening of new markets for conventional herbicides—the adoption of hybrids to the perpetuation of traditional techniques. Electronic devices may reinforce the mechanical orientation of treatment or harvesting operations. By contrast, the contested performances of bovine growth hormone or the interests of feedstuff traders may clash against the heterogeneity of existing livestock.

In the same way, the generalized use of first-generation herbicides helped the development of mechanical techniques, thanks to the simplification of weed control practices. But the diversification of herbicides, the genetic research for new varieties, led eventually to the accumulation and diffusion of a new corpus of biochemical knowledge, agreed upon by the scientific and technological communities, which constitutes the basis for overcoming the mechanical model and formulating a new technical paradigm (Dosi 1982) of agro-food production. Starting from a new generation of herbicides or the creation of a new hybrid, the cultural practices are then redefined, putting into question the use of existing materials or procedures.

Similarly, today the redefinition of the use of chemical inputs, the search for a solution to the problem of residues in food and pollution in the environment, may favor a reorientation of cultivation practices and induce a change of paradigm in the framework of a multipurpose model of agricultural development and space management. Minimum tillage, zero tillage, and strip tillage may lead, for example, to the rehabilitation of mixed crops, rotation, and other techniques of soil conservation. Industrial suppliers able to offer in a *coherent package* new seeds and new techniques (disease and pest control, nutrition, control of informations and agronomic conditions) that reinforce the new orientation of production may gain a competitive advantage in the market. Investments responding initially to an inertial logic may finally induce problems of irreversibility.

Biotechnology as a Flexible Technique in the Post-Fordist Transition

In face of new technical opportunities such as biotechnologies and information technologies, the industrial ability to master the effects of inertia and irreversibility and to dominate the appropriation process of the new techniques (i.e., the appropriation of the quasirent of innovation) is very important. Biotechnologies offer an important opportunity for mastering the transition toward a more diversified production system. Their flexibility allows the opportunity of utilizing them in different forms: either as a temporary solution for problems caused by previous agricultural technologies and as *incremental* innovation in traditional processes (biotransformations, fermentations, breeding of new varieties) or as *radical* innovation, reinforcing the irreversibility of new technical choices.

Because of developments in both its own knowledge bases and their organization, the biochemical industry is best equipped to make use of these specific features of biotechnologies. The industry is in fact perfectly adapted to mass production, but also is capable of diversification in the specialty chemicals and biochemicals. The industry's multisectoral (chemistry, pharmacy, new materials, parapharmacy) and multinational organization and markets allow a simultaneous intervention both in precise segments of the activities—improved seed and hybrids, plant treatment products, veterinary medicine, food additives and preservation products—and in the most artificial and articulated systems of production—greenhouses, irrigated farm systems, confinement housing and feedlot, soilless agriculture. In the short term, the industry can manage in turbulent markets. In the medium and long term, it seems able to master two previously underlined constraints: the redefinition of technical norms (being anchored to the complex and articulated universe of chemistry, biochemistry, and genetics) and the reformulation of price systems (because of the industry's economic dimension and the flexibility of its organization). Over a longer span of time and with a stop-and-go innovation strategy, the multinational corporations may also utilize the new biotechnologies in order to both open up new markets (with biopesticides, for example) in systems that are slow to change because of inertia effects and to consolidate their position through mechanisms of irreversibility. They may then use the appropriation of new techniques as a barrier against competitors, who do not possess the necessary competence and knowledge.

If in the medium term these actors are able to master the evolution of biotechnology, limiting its scope toward specific applications, over a longer span of time they may also consider the possibility of more radical transformations, to avoid displacement by the decline of Fordist production techniques and the emergence of new social functions for agriculture and the rural space. In other words, chemical corporations likely would position themselves to transcend both old modes of agricultural and food production and to withstand environmental politics that would challenge productivist uses of nature. Ultimately, the technological trajectory will be the result of the interactive coevolution of the technical system with other systems (e.g., political, which entails class and power relations, knowledge, norms and values), laying the ground for neo- or post-Fordist structures of production and consumption (Junne 1992).

Conclusions

Economic analyses of biotechnical change in agriculture have been biased by a science-driven orientation. In this essay, underscoring the distance that separates the invention from the innovation, we try to illustrate how numerous intervening economic factors can direct supposed radical innovation toward a conventional path. In the specific case of agrobiotechnology in developed countries, technical discontinuities both between different models and at different points of the food chain, changing patterns of demand for food, the emergence of new functions for rural spaces, new health and environmental norms, and new orientations of the price policy seem more likely to lead to the reopening of technical choices than to the strengthening of the homogenization process.

Biotechnology may play an important role in the process of technical adaptation to industrial flexibility, as the biochemical corporations, because of their strategic position in the agro-food system, their competences, and economic dimension, are identified as important actors in the current transition. Nonetheless, equally important is the role of other social actors and movements (consumers, environmentalists) who are inducing a crisis of the productivist ideology, underlying agro-food development in the postwar era (Buttel 1992).

From a methodological point of view, if our aim is to understand the restructuring process of the agro-food system at the end of the twentieth century, the analysis of the technical sphere is not a sufficient basis for any forecast. In no way can technology be considered an independent variable in the process of social change. Indeed only the cumulative effects of changes in the technical paradigm and in the *social functions* of agriculture and rural space will provide the premise for a "new agricultural evolution" toward models of agro-food development more differentiated and sustainable.

References

Alphandéry, Pierre, Pierre Bitoun, and Yves Dupont. 1989. *Les champs du dèpart.* Paris: Editions La Découverte.

Benvenuti, Bruno. 1990. "Formalisation and the Erosion of Family-Farm Advantages or Else, beyond Mechanicism and Voluntarism." Paper presented at the Fourteenth European Congress of Rural Sociology, Giessen, Germany, July 16–20.

Bonnieux, François, and Pierre Raïnelli. 1988. *Spècialisation règionale de l'agriculture et politique agricole commune.* Rennes, France: Institut National de le Recherche Agronomique-Economie et Sociologie Rurales (INRA-ESR).

Bonny, Sylvie, and Pierre Daucé. 1989. "Les nouvelles technologies en agriculture." *Cahiers d'Economie et Sociologie Rurales* 13(4): 5–31.

Buttel, Frederick H. 1989. "Social Science Research on Biotechnologies and Agriculture: A Critique." *The Rural Sociologist* 9(3): 5–15.

——. 1992. "Ideology and Agricultural Technology in the Late Twentieth Century: Biotechnology as Symbol and Substance." Paper presented at the conference Biotechnology and Agriculture: Technical Evolution or Revolution? Rome, May 28–29.

Byé, Pascal, and Maria Fonte. 1991. "Technical Change in Agriculture and New

Functions for Rural Space." Paper presented at the annual meeting of the American Sociological Association, Cincinnati, Ohio, August 23–27.
——. 1992. "Vers des techniques agricoles fondées sur la science." *Cahiers d'Economie et Sociologie Rurales* nos. 24, 25: 93–114.
Byé, Pascal, Jean-Pierre Frey, and Jean C. Monateri. 1990. "L'innovation phytosanitaire sous le contrôle industriel." Paper presented at the Twelfth World Congress of Sociology, Madrid, July 9–13.
Centre d'Etude pour les Sciences et Les Techniques Avancés (CESTA). 1987. *Sept pistes de rèflexion pour mieux percevoir l'impact des nouvelles technologies du vivant sur l'agriculture et l'agro-alimentaire dans une économie de faible croissance*. Paris: CESTA.
Commission des Communautés Europèennes. 1988. *L'avenir du monde rural*. Brussels: Commission des Communautés Europèennes. Communication de la Commission du 28–7–1988.
Creton, Laurent. 1984. "Les strategies d'innovation progressive." *Revue Française de Gestion* 46: 17–36.
Dosi, Giovanni. 1982. "Technical paradigm and technological trajectories." *Research Policy* 11(3): 147–62.
Dosi, Giovanni, et al. 1988. *Technical Change and Economic Theory*. New York: Pinter Publisher.
Fanfani, Roberto, Raúl Green, and Manuel Rodriguez-Zuñiga. 1992. "Les biotechnologies dans l'agro-alimentaire: un impact limité." *Cahiers d'Economie et Sociologie Rurales* nos. 24, 25: 115–30.
Friedmann, Harriet, and Philip McMichael. 1989. "Agriculture and the State System: The Rise and Decline of National Agricultures, 1870 to the Present." *Sociologia Ruralis* 29(2): 93–117.
Goodman, David, Bernardo Sorj, and John Wilkinson. 1987. *From Farming to Biotechnology*. Oxford: Basil Blackwell.
Green, Keith, and Eduard Yoxen. 1990. "The Greening of European Industry: What Role for Biotechnology?" *Futures* June.
Joly, Pierre-Benoit, and Ehud Zuskovitch. 1990. *L'évolution économique des biotechnologies est-elle prévisible?* Grenoble, France: Institut National de le Recherche Agronomique-Institut de le Recherche Economique et le Planification pour le Developpement (INRA-IREPD).
Joslin, Timothy. 1985. "Marchés et prix: les relations entre l'agriculture et l'économie genèrale." *Economie Rurale* 167: 7–13.
Junne, Gerd. 1992. "Les grandes entreprises face à la revolution biotechnologique." *Cahiers d'Economie et Sociologie Rurales* nos. 24, 25: 143–59.
Kautsky, Karl. 1959. *La questione agraria*. Milan, Italy: Feltrinelli.
Lacroix, Anne. 1981. *Transformation du procés de travail agricole: incidences de l'industrialisation sur le conditions de travail paysannes*. Grenoble, France: Institut National de le Recherche Agronomique-Institut de le Recherche Economique et de Planification pour le Developpement (INRA-IREPD).
Mann, Susan A., and James M. Dickinson. 1978. "Obstacles to the Development of a Capitalist Agriculture." *Journal of Peasant Studies* 5: 466–81.
Mendras, Henri. 1984. *La fin des paysans*. Paris: Actes Sud.
Norgaard, Richard B. 1992. "Sustainability: The Paradigmatic Challenge to Agricultural Economists." In *Sustainable Agricultural Development: The Role of In-*

ternational Cooperation. Twenty-First International Conference of Agricultural Economists, ed. George H. Peters and Bernard F. Stanton, pp. 92–100. Brookfield, Vt.: Dartmouth Publishing Company.

Office of Technological Assessment (OTA). 1986. *Technology, Public Policy, and the Changing Structure of American Agriculture,* OTA-F-285. Washington, D.C.: U.S. Government Printing Office.

Otero, Gerardo. 1991a. "Biotechnology and Economic Restructuring: Toward a New Technological Paradigm in Agriculture?" Paper presented at theannual meeting of the American Sociological Association, Cincinnati, Ohio, August 23–27.

——. 1991b. "The Coming Revolution of Biotechnology: A Critique of Buttel." *Sociological Forum* 6(3): 551–65.

Ruttan, Vernon W. 1990. "Resource and Environmental Constraints on Sustainable Growth in Agricultural Production: Report on a Dialogue." Staff paper, Department of Agricultural and Applied Economics, University of Minnesota, Minneapolis.

Servolin, Claude. 1972. "L'absorbtion de l'agriculture dans le mode de production capitaliste." In *L'univers politique des paysans,* pp. 41–77. Paris: A. Colin.

Sigaut, François. 1989. "La Chine, l'Europe et les techniques agricoles." (Note critique.) *Annales ESC* 1: 207–16.

Van der Ploeg, Jan. 1990. *Lo sviluppo tecnologico in agricoltura: il caso della zootecnia.* Bologna, Italy: Il Mulino.

Weber, Florence, and Sylvain Maresca, eds. 1986. "Travaux et mètiers: la confusion des activitès en milieu rural." *Cahiers d'Economie et Sociologie Rurales* 3 (Special issue).

Wilkinson, John. 1993. "S'adapter à la demande alimentaire: nouvelle orientations industrielles en matière d'innovation." *Cahiers d'Economie et Sociologie Rurales* nos. 24, 25: 131–42.

11

Distance and Durability: Shaky Foundations of the World Food Economy

Harriet Friedmann

The first outbreak of the food crisis in the early 1970s was as devastating for the Third World as the energy crisis. The dramatic changes now under way have roots in the first economic breaches of the Cold War dam. Although they were anticipated by smaller transactions in the late 1960s, the Soviet-U.S. grain deals of 1972–73 were so large that they precipitated a prolonged, still unresolved, crisis of the postwar food regime. Since the Soviet-American grain deals were the economic expression of Detente, it is now clear that the stable, if unequal, relations of the postwar food regime were bound up with the mutually exclusive trading blocs of the Cold War. The blocs had provided the framework for decolonization and building of national economies, including food and agriculture, in the Third World. The crisis of the food regime has been bound up with a restructuring of the framework of rival blocs, which began not in 1989, but two decades earlier.

The postwar food regime consisted of distinct complexes. The most important changes in the food regime can be traced through the wheat complex, the durable food complex, and the livestock complex. Each complex is defined as a chain (or web) of production and consumption relations, linking farmers and farm workers to consuming individuals, households, and communities. Within each web are private and state institutions that buy, sell, provide inputs, process, transport, distribute, and finance each link. Each complex includes many class, gender, and cultural relations, within a specific (changing) international division of labor. Each evolved within the politically bounded economic space of the West, until the major transactions of the early 1970s irrevocably linked the Cold War blocs, leading finally to the dissolution of the bloc structure itself.[1] Since the early 1970s the implosive merging of the blocs had coincided with a greater intertwining of the complexes.

Although each country or region inherits a specific legacy of incorporation and marginalization, each of the three complexes created general conditions for the Third World. The wheat complex facilitated food import dependency. The durable food complex reduced demand for traditional tropical exports, especially sugar and vegetable oils. The livestock complex (like fruits and vegetables) shifted from a national to a transnational basis; by taking hold directly of production, it differentiates the Third World in new ways and, more directly than the wheat complex, undermines local, mixed economies.

The Wheat Complex

For the Third World the wheat complex was the major source of food import dependency.[2] When previously self-provisioning countries began to import food in the 1950s and 1960s, the food they imported most was wheat, no matter what the customary dietary staples. As a result, when they became dependent on imports they were hooked on the most expensive grain. The average price per ton of wheat in 1959–61 was US$66, 25 percent higher than maize, more than six times higher than rice (Rowe 1965:5), and almost five times the price per ton of petroleum.

Why should importers buy such expensive grain, particularly in those countries in which it was not a customary food? The answer is that much of the wheat trade was organized not through markets, but through other agreements.[3] Third World countries, both colonies gaining independence and Latin American countries that had pursued import substitution policies in the 1930s and 1940s, had little foreign exchange, and that little had to buy whatever imports were necessary to build industries.

At the same time, wheat was available in a very particular way. At 39.5 million metric tons in the late 1950s, it was second in volume only

[1] A separate story must be told about the state socialist bloc and the Third World countries that joined it. For commentary on some of these cases, see Wadekin 1990.

[2] For more extended discussion, see Friedmann 1982, 1990b. Unless otherwise indicated, all data and arguments are based on the former article.

[3] The effect on the residual market was chronic oversupply. Even when U.S. grain was not transferred through aid but sold commercially, the price was depressed by the nearly 40 percent of world trade that took the form of aid. See Friedmann 1982.

to petroleum among primary commodities traded, and was many times higher than rice and maize (Rowe 1965). More important, the United States, whose historical share of the world wheat market declined drastically during the depression, was accumulating large public stocks of surplus wheat after World War II through New Deal price support programs. These stocks created supplies available for export in quantities greatly exceeding demand.

Third World wheat imports came overwhelmingly from the United States, and at the outset were often subsidized heavily through "concessional sales" in nonconvertible currencies at negotiated prices, under Public Law 480 (food aid). Based on a new axis between the United States and the Third World, between the early 1950s and the early 1970s, world wheat exports increased 2.5 times. In the 1950s and 1960s the U.S. share of world wheat exports grew from just over a third to more than half. And as Europe substituted domestic production for its historic imports, the Third World and Japan became the major importers. The Third World share of wheat imports grew from 19 percent in the late 1950s to 66 percent in the late 1960s (Friedmann 1982, 1990b).

At the end of World War II, no Third World country had been a significant importer, and some, especially in North Africa and Latin America, had been major or minor exporters (Argentina and Brazil still are). Between the early 1950s and the late 1970s, per capita consumption of wheat increased by 63 percent in the market economies of the Third World and not at all in the advanced capitalist countries. By contrast, per capita consumption of all cereals except wheat in the Third World increased only 20 percent, and per capita consumption of root crops actually declined by more than 20 percent.[4]

Why did Third World countries import wheat instead of investing in national food supplies to support industrial and urban development, as European countries had done earlier? The answer is both material and ideological. First, the United States wanted to get rid of the surplus stocks of wheat it had accumulated through domestic farm programs, and exporting wheat to Third World countries conveniently coincided with a mix of foreign policy and humanitarian goals. The U.S. government, through the Marshall Plan, had invented foreign aid as a mechanism to increase trade despite lack of dollars by prospective importers. In 1954 it adapted this mechanism to food aid through P.L. 480. Sec-

[4] Computed from FAO production and trade data. The early 1950s refers to the five-year average 1948–52, and the late 1970s refers to the five-year average 1976–80.

ond, Third World countries welcomed cheap food, that is, wheat imports subsidized by the U.S. government, as an aid to creating an urban working class. The ideology of development focused, to the point of obsession, on industrialization.

By the early 1960s U.S. food aid accounted for 35.6 percent of total world wheat trade. As a result, in Colombia, for example, the total area cultivated for commercial foods fell by 54 percent, and imported wheat came to account for 78 percent of domestic wheat consumption (Dudley and Sandilands 1975). By the late 1960s the aim of creating commercial markets through aid had begun to succeed. Many Third World countries depended on wheat imports. But two new factors undercut the dependence on U.S. wheat. First, other advanced capitalist countries, notably the European Community, both accumulated surpluses through their own farm programs and were strong enough to compete in the aid game. When the United States went into trade deficit, leading to the first dollar crisis, it lost some of its advantage over competitors in concessional sales. Second, the Green Revolution began to show results for import substitution, for example, in India, a major importer in the mid-1960s.[5]

In the early 1970s the structural contradictions of the wheat complex came violently to a head with the crumbling of the Cold War economic dam. Soviet purchases from the United States in 1972–73 were on a scale that overshadowed all other transactions combined. Although other temporary factors also contributed, the Soviet purchases created sudden shortages in a market based on chronic surpluses. Prices tripled and quadrupled, and food aid plummeted. The dilemma of many Third World countries was that they had gotten hooked on imports in the 1950s and 1960s, and now had to buy at high prices, just when oil prices soared, too.

By the late 1970s surpluses returned, but another oil price rise, the debt crisis, and fierce export competition prevented a return to the old regime. The response to temporary shortages (and high prices) had been a credit-based expansion of production by U.S. (and other) farmers and a credit-based shift from subsidized to commercial consumption by Third World states. This double debt exacerbated the problem of unsustainable, competitive dumping of chronic surpluses.

World trade in wheat dropped sharply in the mid-1980s, and the U.S.

[5] The Green Revolution can be seen as the export of a specific model of agriculture based on hybrid seeds and intensive use of chemical and mechanical inputs, introduced in the United States with hybrid maize in the 1930s and 1940s. See Kloppenburg 1984 and Philip Raikes's analysis of Tanzania (1989).

share fell even more, from 43 million metric tons to 29 million between 1980 and 1985 (USDA 1986:11). Impending trade wars led to struggles over whether, how, and how quickly to include agricultural products in the GATT. Despite implicitly coordinated attempts by major exporters to reduce domestic grain surpluses, export subsidy wars continued, creating a prolonged impasse in the Uruguay Round (Friedmann 1993b).[6]

Then, at least until collapse of the USSR, incorporation of the socialist bloc became the major terrain of competitive struggle within the wheat complex (van der Pijl 1984; Friedmann, 1990c, 1991b). As the prospect for subsidized imports faded, many Third World countries faced the dismal dilemma of chronic debt and food import dependence. Structural adjustment (austerity) measures as conditions for rescheduling debt invariably restrain imports and reduce or abolish food price subsidies to consumers. The only bright light of rising food prices might be encouragement of domestic agriculture after decades of devastation by subsidized imports in the wheat complex. Yet structural adjustment also forces domestic agriculture into export production to pay debts, and privatization of public projects often compromises existing potential.[7]

The wheat complex is losing its distinct existence. Politically, it was based on price support programs that generated surpluses in advanced capitalist countries. These policies were based in turn on the political strength of farm lobbies, yet divided those lobbies into commodity-specific interest groups. They also encouraged drastic concentration and reduction of the numbers of farmers.[8] Potentially new forms of

[6] The United States had insisted on their exclusion from GATT at the outset because it wished to retain domestic farm subsidies. This created conditions conducive to similar farm programs in Europe, Japan, and elsewhere.

[7] For an analysis of the mainly negative results for domestic food production of one case of privatization, a state-irrigated rice project in Senegal, see Woodhouse and Ndiaye 1990. Zimbabwe's experience of food price increases shows different but equally bad results for domestic food supplies. After imposition of a World Bank/IMF stabilization program in 1982/83, the price of maize meal rose 73 percent and wages were frozen for almost two years. Although "incentive" prices led to national surpluses of maize, the apparent success was based on drawing more from the communal sector, which had supplied 4 percent of the maize to the state marketing board before the measures and 37 percent afterward—at the expense of nutrition in these areas. See Amin 1989. Since IMF policies were imposed in 1983 (and voluntarily continued by the government since 1985), Brazil has become a major competitor in world markets—the leading new agricultural country. Yet its per capita supply of foods has plummeted, because of both slashed wheat imports and decimated domestic production of food staples (George 1988:145–47, and de Melo 1986).

[8] For an overview of the U.S. farm crisis, see Buttel 1989.

agricultural regulation, based on urban political support, are emerging both in advanced capitalist countries and in the Third World.[9]

Economically, the distinct identity of the giant grain companies had been tied to family wheat farms,[10] as well as the growth of subsidized U.S. (and later other) exports (Morgan 1979:vii–xiv). Now that new markets for wheat will come mainly from new uses, the anachronistic mercantile companies have speeded up their diversification and adoption of modern industrial practices, linking themselves to the durable food and livestock complexes (Kneen 1990, Morgan 1989).

For the state system, the surpluses consistently reproduced within the capitalist bloc flowed into the world economic sea created by the irreversible Soviet entry into the complex as a buyer. Prices were out of reach for indebted Third World importers, despite the intensely competitive export environment. Some new agricultural countries (NAC), especially Argentina and Brazil, quickly entered the fray. The collapse of socialist markets then brought further anarchy to world markets.

The Durable Food Complex

The durable food complex contributed to decining demand for traditional (tropical) exports of the Third World. During the 1950s and 1960s food manufacturing corporations in advanced capitalist countries increasingly replaced simple refiners and final consumers as buyers of tropical agricultural products. Corporate research departments began a systematic search for chemical and biological substitutes for industrial raw materials, including the key ingredients of sweeteners and fats. Sugar and vegetable oils such as peanut, palm, and coconut, among the most important Third World export crops, lost their climatic monopoly and became one of a growing category of substitutable raw materials for industrial foods. Since temperate crops were efficiently produced, often subsidized, and accessible to direct contracts assuring predictable supplies, industrial substitution of tropical products was simultaneously import substitution of tropical exports (Friedmann 1991a).

[9] Two extreme experiments are New Zealand, which has abolished subsidies completely, and Sweden, which has adopted a very interesting new type of regulation based on ecological and food quality issues, among others. See Cloke 1989 and Vail 1987. EC membership will change this. Regional food economies may have more promise (Friedman 1993a).

[10] The family farm has lost its content, and family wheat farmers have lost their political strength. See Friedmann 1990a and Strange 1988.

Accompanying the industrialization of food are changes that David Goodman, Bernardo Sorj, and John Wilkinson call "appropriation" and "substitution" (1987). As agricultural input industries succeed in reducing the vagaries of nature in one aspect after another of agriculture, they "appropriate" part of the production process from farmers. As food manufacturing industries succeed in replicating the natural properties specific to traditional ingredients, they "substitute" the products of farmers.

Appropriation and substitution have occurred at specific places within the world economy and state system. Appropriation was much more important to the industrial agriculture of advanced capitalism, at least during the food regime of the 1950s and 1960s, and substitution had its greatest impact on the Third World.

Appropriation represents much deeper control than the monopoly of markets in the wheat complex. The humble potato, classically the cheapest food of the industrial working classes of Europe and North America (Salaman 1949), is an example of the simultaneous transformation of diets and appropriation of agriculture. The potato became the basis for one of the largest corporate empires in the world, McCain, whose first major product was frozen french fries, followed by other new high value-added foods. To assure a steady supply of genetically standard fresh crops, McCain reorganized traditional agricultural communities in eastern Canada. Monopoly contracts specifying most aspects of production subordinated family farms and created a monocultural region (Murphy 1987). A greater degree of appropriation occurred when giants such as Bud Antle expanded to supply year-round consumption of crops, such as lettuce and tomatoes, which were bred for machine harvesting and long-distance shipping. Such continuity of supply allowed the development of monocultural, industrial farming in the western United States, with an integrated labor process using paid, often unionized, labor (Friedland, Barton, and Thomas 1981, Thomas 1985).

Substitution was far more important for the fate of tropical exports and the Third World countries that dependend on them. Industrial chemists made new sweeteners from grain and improved the blending and substitution of oils. In advanced capitalist countries temperate oils and grains were efficiently produced, and various farm programs kept supplies high. It is not surprising, therefore, that despite increased requirements for sweeteners and oils in manufactured foods, the terms of trade for tropical exports declined in the 1950s and 1960s.

Sugar substitution moved from the field to the laboratory with the development of high fructose corn syrup. Import substitution with beet sugar, which goes back to the blockades of Napoleonic Europe, was a protectionist measure in both Europe and the United States (which during the food regime of the fifties and sixties also subsidized domestic cane production). High fructose corn syrup and other industrially produced sweeteners, however, created the possibility of using all grains as raw materials.

When the price of sugar soared in the early crisis years 1974 and 1975, the relative price of surplus grain declined, and with it the cost advantages of producing high fructose corn syrup. Although the cost differential quickly disappeared with the collapse of world sugar prices later in the decade, supplies of domestic corn—and with scientific advances, other grains that could be converted to sweeteners—were more stable as sugar prices became increasingly volatile.[11] By the mid-1980s new sweeteners accounted for more than 40 percent of U.S. consumption, and U.S. per capita consumption of sugar had declined by 30 percent (Mahler 1986).

Despite a production increase of two-thirds, world sugar trade actually fell 3.5 percent from 1960 to 1988. At that time, Third World countries as a group imported more sugar than advanced capitalist ones. The earning power of sugar became completely unreliable, yet six poor countries were so vulnerable to these fluctuating prices that their export earnings declined by one-third or more with a 1 percent fall in the world price of sugar.[12] An exceptional success in cane sugar was Brazil, with a new twist on substitution: using sugar-derived fuel as a substitute for energy imports.

The substitution of tropical oils owes less to the laboratory and more to the complex politics and economics of the livestock sector. Although palm, coconut, peanut, and cotton oils always competed with temperate oils, such as sunflower, canola (rapeseed), and mustard, the major change of the postwar era pivoted on a new temperate oil. Soya oil was originally developed before the war, with U.S. state subsidies, to supply the margarine industry (Bertrand, Laurent, and LeClerq 1983). After the war it became more important as a joint product of soya meal, produced by high technology industries for the intensive livestock industry.

Soybeans in the 1950s and 1960s came to account for one-third of

[11] With an index of 100 in 1980 (the highest price between 1975 and 1988), the price of sugar was 33.65 in 1979 and 58.77 in 1981. The lowest in the period was an index of 14.22 in 1985 (FAO 1988).

[12] In order of vulnerability they are Guyana, Cuba, Jamaica, Dominican Republic, Mauritius, and the Philippines (FAO 1984a).

world oilseed production, and before the food crisis the United States accounted for more than 90 percent of exports. Soya oil was part of Marshall aid to Europe, and was second to wheat in P.L. 480 shipments. Eventually Mediterranean countries began to import cheap soya oil and export expensive olive oil. In the Third World, three of the four major recipients of soya oil under P.L. 480 became major commercial importers.[13]

With the Soviet purchases of the early 1970s prices soared, and the U.S. government feared domestic shortages. It placed a temporary embargo on soybean exports. Although all contracts were ultimately honored, this embargo marked the end of U.S. dominance of the world market. Brazil and Argentina seized the opportunity to cut into world exports, becoming important NACs. As a result domestic soya oil, a byproduct of the highly regulated soymeal industry, supplies 90 percent of the Brazilian market (LeClerq 1989).

The possibilities for substitution in the durable food complex have serious implications for the debt-inspired export push that began in the 1980s. Sugar and vegetable oils are increasingly substitutable industrial raw materials, rather than final consumer goods. Corporations that invent, market, and buy inputs for increasingly complex edible commodities went through an intense period of concentration in the seventies and eighties. The 1987 takeover of Beatrice Foods, which began in the 1890s as a dairy in Beatrice, Nebraska, then the largest transaction in history (Sterngold 1987), was topped by the merger of Kraft and General Foods in 1989 (Davis 1989:30). As fast-food corporations such as McDonald's expand into Third World and former socialist markets, the shift to industrial raw materials is accelerated further. Substitution and corporate concentration mean that efforts to increase tropical exports are doomed to a vicious spiral of declining prices.

The turn to nontraditional exports creates new problems. Fresh fruits, vegetables, fish, even flowers and ornamental plants, often compete directly with domestic products in the United States and the EC. Though they are often based on national capital (for instance in Chile) they are organized, from seeds to marketing, by global standards (Gomez and Goldfrank 1990). In the shift from colonial complementarity to competitiveness within a global sector, the political basis of produc-

[13] These data are based on FAO and U.S. Department of Agriculture publications.

tion and trade does not go away, but it does change. This fact is clearly seen in the livestock complex.

The Livestock Complex

The livestock complex parallels industrial restructuring in the world economy. During the food regime of the 1950s and 1960s animal and crop production were reorganized, first in the United States and then in Europe. Although intensive, specialized livestock production was national, new feed crops were, from the outset, industrial raw materials for the transnational feed manufacturing industry. After the crisis began in the early 1970s the complex exploded into intense international competition. On the side was fierce competition between the United States and the EC, linked to the conflict within the wheat complex. On the other was the competition in increasingly substitutable feed crops from the NACs—parallel to newly industrialized countries (NICs).[14] This ended irrevocably the division between First and Third World exports, and at the same time led to the differentiation within the Third World.

High meat consumption was key to the postwar diet of the United States and Europe. The demand was supplied by a new specialization into two distinct subsectors—intensive, large-scale, often industrially organized livestock operations and monocultural production of feed crops.[15] This pattern began in the United States and was founded on three key factors: a revolution in maize production based on hybrids requiring intensive mechanical and chemical inputs; the massive introduction of an Asian plant, soya, which substituted a commercial feed crop for forage crops (forage became redundant as tractors replaced draft animals); and a new capital-intensive feedstuffs industry that interposed itself between crop and livestock producers and organized

[14] Like fresh fruits, vegetables, and ornamental plants, exports of inputs to the feedstuffs industry compete successfully with advanced capitalist production.

[15] Dairy, the most subsidized and controlled subsector of advanced capitalist agriculture, adapted new techniques with considerably less specialization or social transformation. It also has produced the most embarrassing surpluses. Attempts to reduce surpluses threaten the framework that has allowed dairy farms to increase productivity, scale, and even upstream and downstream dependence on agri-food corporations, within the traditional framework of mixed family farms. For a comparative analysis of capitalist dairies in California and family dairies in Wisconsin, see Gilbert and Akor 1988.

both sectors through long-term contracts (Kloppenberg 1988, Friedman 1991a, Kenny et al. 1991).

Soya and maize were as important to the emergence of the livestock complex as factory production of poultry and pork and the growth of cattle feedlots. Soybeans are by far the fastest-growing crop in world agriculture since 1945. From an Asian food crop, soybeans became the basis for a global transformation of livestock production, linking field crops with intensive, scientific animal production, through giant agri-food corporations, across many national boundaries. World production of soybeans tripled between the late 1940s and 1970, and then doubled again between 1970 and 1988, virtually all in the West. At the end of the 1980s more than a quarter of world production, worth US$4.8 billion entered international trade. At the same time maize, whose production was totally transformed via specialized, capital-intensive use of hybrids, was worth $75 billion in international trade, six times the value of the world wheat trade (FAO 1987, 1988).

During the food regime, intensive livestock production was a preeminently national sector, first in the United States, then in Europe and other advanced capitalist countries. Yet it shared with industries of the period a deeper process of transnational, intrasectoral integration (Aglietta 1982). Like automobiles or aircraft—in which multiple components produced in different factories and in different national economies came to be linked by transnational corporations through either direct subsidiaries or subcontracts—the specialized livestock sector was connected, via the transnational feedstuffs industry, to specialized crop farmers.

Farm level specialization by no means automatically led to U.S. domination of world feed crop production. Once crop and livestock producers were linked by corporations, inputs could in principle come from anywhere, including the Third World. The major processors, represented by the powerful American Soybean Association, were American based, including the largest merchant grain companies, Continental and Cargill.[16] When they established operations in Europe after World War II, Continental and Cargill were joined by other transnational corporations, notably the Anglo-Dutch company Unilever and the Argentine-based grain trading company Bunge and Born. Unilever was a pioneer

[16] The others were Archer Daniel Midlands, Anderson Clayton, Central Soya, Ralston Purina, and A. E. Staley. See Bertrand, Laurent, and LeClerq 1983. With the concentration of agri-food industries in the 1970s and 1980s, some nationalities changed too. For instance, Mitsui of Japan became a major U.S. actor when it bought bankrupt Cook Industries in 1978. See Kneen 1990.

in global sourcing of raw materials, including tropical oilseeds from Asia and Africa, as well as domestic European oilseeds and grains,[17] and Bunge and Born had access to linseed and other South American oilseeds.[18]

Yet U.S.-European negotiations led the new European Economic Community to exempt soybeans from import duties, in return for which the United States accepted EEC protection against wheat.[19] From the U.S. side, European imports of maize and soya more than compensated for lost wheat sales (Morgan 1979). For Europe soya represented a "spearhead of American agricultural expansionism" (Bertrand, Laurent, and LeClerq 1983:5).

Though soya has definite advantages over other oilseeds,[20] Jean-Pierre Bertrand, Catherine Laurent, and Vincent LeClerq argue that a complex politics surrounded its "victorious battle for the American market," followed by its "conquest of the world" (1983:50–59). Each conquest followed two phases. First was the export of soya oil through P.L. 480, which accounted for almost three-quarters of U.S. vegetable oil exports in the early years and made soymeal available as a byproduct. Next came the international agreements that made it cost-effective for European feedstuffs manufacturers to substitute imports of U.S. soybeans and maize for other oilseeds and feedgrains. The mass consumption of livestock products, which was crucial to the new diet of postwar prosperity, was to be supplied through the American system of intensive livestock production based on U.S. raw materials (Bertrand, Laurent, and LeClerq 1983:50–59).

In 1972–73 soybeans and maize were as implicated as wheat in the Soviet purchases that precipitated the long-term crisis of the food re-

[17] "Unilever's Sticky Fingers: Inside a Multinational," *New Internationalist,* June 1987: 22–23. Of course, European farmers had long produced oilseeds and coarse grains for animal feeds as well as other uses. See Goodman, Sorj, and Wilkinson 1987.

[18] Despite worldwide diversification into numerous product lines and a shift of headquarters to Brazil, Bunge and Born retained such a powerful place in the Argentine economy that it wrote the economic policy for the former Argentine government (Morgan 1989:46–48).

[19] Grain company executives were influential in putting pressure on the Common Agricultural Policy through the U.S. Department of Agriculture, though this was attenuated by the larger U.S. interest in promoting this first major step toward European integration (Morgan 1979:128–34). Soya virtually wiped out other oilcakes based on domestic or colonial crops, such as flax or cotton. By 1981 industrial soycakes, still using largely imported soybeans, had leapt from 7 percent to 86 percent of French production (Bertrand, Laurent, and LeClerq 1983:63).

[20] Soya is an excellent mealcake because it contains all the amino acids usually found in animal products. Maize was mixed with the soymeal cakes so manufacturers could evade the maximum prices on soy products set by the U.S. government in encouraging intensive livestock production (Bertrand, Laurent, and LeClerq 1983:45–49).

gime. Panicked by the prospect of domestic shortages during the scandals surrounding the companies involved in the sales, the U.S. government embargoed soybean exports. Although all contracts were finally met, Japan quickly sought to diversify supplies, and other countries, notably Argentina and Brazil, seized the opportunity to enter a market with (temporary) shortages and high prices. The U.S. share of soya exports dropped from a near monopoly to about two-thirds within a very short time.[21]

Soon there were two types of NAC. One type entered world export markets of inputs to the feedstuffs industry. In a second type of NAC, especially the newly industrialized countries and the oil-rich countries, states intervened in agriculture and trade to create a domestic intensive livestock industry, using industrial feedstuffs.[22] In Brazil the two were combined and integrated through strong state organization of a national sector centered on a domestic soymeal industry.[23]

The restructuring of the livestock complex has changed international relations by tying it more closely to the durable foods and wheat complexes. For example, Brazil exports pulp from the frozen orange juice industry and Thailand exports tapioca as inputs to the feed industry. These exports, and other grains, challenge the virtual monopoly of hybrid maize and soya as the energy and protein components of composite feedstuffs, further undermining the dominance of the United States. Consequently, the political factors so central to the rise of the livestock complex—price supports, export subsidies, and international agreements, notably the exclusion of agriculture from the GATT[24] no longer work to the advantage of the United States. Brazilian exports of soya

[21] Brazilian production leapt from 1.5 million to 13.5 million metric tons in the seventies, and then to 18 million in 1988. Another NAC, Argentina, expanded soybean production in the eighties, reaching 9.8 million metric tons in 1988. By contrast U.S. production fell somewhat to 41.8 million metric tons, from almost two-thirds to under half of world production during the 1980s. Meanwhile the People's Republic of China produced consistently about 10 million metric tons over the whole period, but in the eighties Asian NACs added 5 million metric tons to world production (FAO Yearbook, various years).

[22] Countries with high domestic subsidies for poultry or pork include Brazil, Taiwan, and South Korea. Mexico has very high subsidies for soybeans (Miner and Hathaway 1988:51).

[23] Contrary to the old colonial pattern, Brazil dominates world soymeal exports, and U.S. exports are overwhelmingly of unprocessed beans.

[24] Although it is rarely noted in press coverage of the conflicts between the United States and the EC over inclusion of agriculture in the GATT, the United States originally insisted on the exclusion so it could legally continue its domestic farm support programs. The protection of EC domestic markets is different from U.S. policies, but like export subsidies, each is linked to their specific histories as exporters or importers as the framework for farm price support policies.

meal, fostered by a government-supported national soya-milling (and poultry) industry, had a higher unit value than U.S. exports, which were mostly unprocessed soybeans (LeClerq 1989).

The emergence of a world sector, with global sourcing of feed inputs and global marketing of meat-related commodities, has equally important class dimensions. Like the wheat complex before the crisis, the incorporation of Third World countries into the livestock complex since the early 1970s intensified the pressures on mixed, self-provisioning agriculture (and related local markets). But where the wheat complex undermined local production through imports, the livestock complex takes hold directly of production in some places. In all places it marginalizes traditional production and, with it, access to a complex range of products.

Steven Sanderson argues for a tendency to create a "world steer" parallel to the better-known "world car." This is a high-quality product, standard from its genetic lines to its packaging. Regardless of the nationality of ownership, the world steer reorganizes beef production to meet international (for Mexico, U.S., and Japan) standards through expensive feeds and medicines, concentrated feedlots, and centralized slaughtering. The displacement of traditional marketing and processing means that small sideline producers lose access to markets. Without this source of income, they often find it difficult to continue to raise cattle even for self-provisioning. As a result, they lose milk and meat. Less obviously perhaps they also lose leather and tallow, and thus local supplies of clothing, footwear, and cooking fat. In Mexico in the 1980s the standard of living and security of peasant communities fell as the national accounts suffered the import of leather (Sanderson 1986).

With the commodification of diets and greater inequality of incomes, it is possible to see a division of the livestock complex to serve rich and poor consumers—remembering that many are too poor to buy at all. Whereas the world steer reflects expensive standards, such as marbling, for elite markets, beef that enters the mass markets of the durable food complex seeks cheap land and labor. Extensive grazing in the Amazon region and in Central America produces beef for the low-quality market, destined for hamburgers, frankfurters, and complex food products containing beef as an ingredient. This has devastating ecological as well as social effects (Skinner 1985, Feder 1980).

Finally, the circle closes when maize, tapioca, and all other staple foods become export crops for use in animal feeds or in the production of generic ingredients, such as starches and sweeteners. It not only pits humans against animals for consumption of traditional staples,[25] but seizes all crops as raw materials for complex edible commodities passing through corporate production and distribution.

Conclusions

The wheat, durable food, and livestock complexes, plus others such as fruits and vegetables, are becoming increasingly intertwined. This brings together the contradictory tendencies of each complex. The crisis that began with detente is deepening as the countries of the disintegrating Soviet bloc seek incorporation into the capitalist world economy and state system. It is clear that the past cannot return. What are the current possibilities?

The dominant tendency is toward *distance* and *durability*, the suppression of particularities of time and place in both agriculture and diets. More rapidly and deeply than before, transnational agri-food capitals disconnect production from consumption and relink them through buying and selling. They have created an integrated productive sector of the world economy (Hoogvelt 1982), and peoples of the Third World have been incorporated or marginalized—often simultaneously—as consumers and workers.

If this reached its logical end, consumers of corporate food products would be differentiated by class, rather than nation or cultural region. Farms would adapt production to demand for raw materials by a small set of transnational corporations (not a fancied "world market") and in order to meet quality standards would buy inputs and services from (often the same) transnational corporations. Relations of consumption and production would be shaped (large or small, with specific gender, age, and class characteristics) within the possibilities of incorporation or marginalization by agri-food complexes.[26]

Paradoxically, as industrialization unifies agriculture, it blurs its boundaries.[27] To ignore this reality for the convenience of national and

[25] The point has been made in different ways that rarely stress the capital-intensive character of the feedstuffs industry.

[26] For an exemplary analysis of these trends, and much information on the secretive giant, Cargill, see Kneen 1990.

[27] This point makes it even more crucial to draw out the implications of Henry Bernstein's important political insight into the "modernisation" model: that it collapses the distinction between the conditions of the agricultural sector and the needs and capacities of farmers. The former, which concerns technical functions and price relationships, is privileged, even when the latter is addressed in some way. Although Bernstein is discussing one country, Tanzania,

sectoral statistics is to miss the opportunity to understand, for example, the connections among extensive cattle production in the Amazon, intensive cattle production in Mexico, canned beef in Mexican border industries, frozen hamburger patties and boxed prime beef in supermarket freezers throughout the world, and McDonald's in Budapest and Hong Kong. These intertwined "beef chains" intersect with parallel "potato chains" and many other chains of inputs to complex final products, which are somewhat arbitrarily labeled "agricultural" (irradiated potatoes), "industrial" (frozen fries), or "service" (hot fries).

At the same time the role of states and the state system has paradoxically provided supports for the transnational agro-food sector. After World War II state agricultural and trade policies, within the framework of monetary rules and military alliances, created an institutional framework for fledgling agri-food capitals to become strategic poles of accumulation and to integrate transnationally the increasingly specialized subsectors of the food regime. There were other alternatives, based on subordinating international trade and investment to national planning or—in the narrowly defeated World Food Board proposal of 1947—to subordinate national agriculture to a powerful international planning agency.

There are alternatives to the rapid but unstable homogenization of the world's agriculture and diets. As capital dissolves the boundaries between sectors, it creates more than one possibility for relinking producers and consumers. Political strategies can build on social, cultural, and economic pressures to relink them locally.

Three political movements provide the basis for this possibility: regional autonomy (which paradoxically reinforces corporate transnationalism in undermining national economies), the creation of multinational (or transnational) regions, and the shift from rural to urban populations. The new agricultural countries, like the newly industrialized countries (they are sometimes the same), like the very poor countries of Africa, and like the countries of Europe and North America, show complicated tensions between transnational, national, and—though they are more difficult to trace statistically—regional levels of ownership and regulation. The pressures to regionally reconstruct links between producers and consumers is apparent in many places, whether from economic desperation or from urban politics that place a higher

---

the gap between the conditions of agri-food accumulation and the needs of farmers (and consumers) is still greater when mapped onto a transnational set of agri-food relations (Bernstein 1989:17).

priority on ecologically sound land use and uncontaminated foods than on the social and technical imperatives of monocultural farming.

Capital has undermined the traditional integrity of agriculture and local diets everywhere. The past cannot endure; the future is open.

References

Aglietta, Michel. 1982. "World Capitalism in the Eighties." *New Left Review* 136: 5–42.

Amin, Nick. 1989. "Maize Production, Distribution Policy, and the Problem of Food Security in Zimbabwe's Communal Areas." Working Paper no. 11, Development Policy and Practice Working Group, Open University, Milton Keynes, England.

Bernstein, Henry. 1989. "Agricultural 'Modernisation' in the Era of Structural Adjustment." Working Paper no. 16, Development Policy and Practice Group, Faculty of Technology, Open University, England.

Bertrand, Jean-Pierre, Catherine Laurent, and Vincent LeClerq. 1983. *Le monde du soja*. Paris: La Decouverte.

Buttel, Frederick H. 1989. "The U.S. Farm Crisis and the Restructuring of American Agriculture: Domestic and International Dimensions." In *The International Farm Crisis*, ed. David Goodman and Michael Redclift, pp. 46–83. London: Macmillan.

Cloke, P. 1989. "State Deregulation and New Zealand's Agricultural Sector." *Sociologia Ruralis* 29(1): 34–48.

Davis, L. J. 1989. "Philip Morris's Big Bite." *New York Times Magazine*, April 9.

de Melo, F. Homem. 1986. "Unbalanced Technological Change and Income Disparity in a Semi-Open Economy: The Case of Brazil." In *Food, the State, and International Political Economy*, ed. F. L. Tullis and W. L. Hollist. Lincoln: University of Nebraska Press.

Dudley, Leonard, and Roger J. Sandilands. 1975. "The Side Effects of Foreign Aid: The Case of Public Law 480 Wheat in Colombia." *Economic Development and Cultural Change* 23(2): 325–36.

FAO. 1984a. *Promoting Agricultural Trade among Developing Countries*. Economic and Social Development Paper no. 41. Rome: FAO.

——. 1984b. *Quarterly Bulletin of Statistics*. Various issues.

——. 1987. *Trade Yearbook*. Rome: FAO.

——. 1988. *Production Yearbook*. Rome: FAO.

Feder, Ernest. 1980. "The Odious Competition between Man and Animal over Agricultural Resources in the Underdeveloped Countries." *Review* (Fernand Braudel Center) 3(2): 463–500.

Friedland, William H., Amy E. Barton, and Robert J. Thomas. 1981. *Manufacturing Green Gold: Capital, Labor, and Technology in the Lettuce Industry*. Cambridge: Cambridge University Press.

Friedmann, Harriet. 1982. "The Political Economy of Food: The Rise and Fall of the Postwar International Food Order." In *Marxist Inquiries*, ed. Michael Burawoy and Theda Skocpol. *American Journal of Sociology* 88: S248–82.

—-. 1990a. "Family Wheat Farms and Third World Diets: A Paradoxical Relationship between Unwaged and Waged Labor." In *Work without Wages: Comparative Studies of Housework and Petty Commodity Production,* ed. Jane L. Collins and Martha E. Gimenez. Albany: State University of New York Press.
——. 1990b. "Origins of Third World Dependence." In *The Food Question,* ed. Henry Bernstein, Ben Crow, Maureen Mackintosh, and Charlotte Martin. London: Earthscan.
——. 1990c. "Rethinking Capitalism and Hierarchy." *Review* (Fernand Braudel Center) 13(2): 255–63.
——. 1991a. "Changes in the International Division of Labor: Agri-Food Complexes and Export Agriculture." In *Towards a New Political Economy of Agriculture,* ed. William Friedland, Lawrence Busch, Frederick Buttel, and Alan Rudy. Boulder, Colo.: Westview Press.
——. 1991b. "New Wines, New Bottles: Regulating Capital on a Global Scale." *Studies in Political Economy* 36: 9–42.
——. 1993a. "After Midas' Feast: Alternative Food Regimes for the Future." In *Food for the Future: Conditions and Contradictions of Sustainability,* ed. Patricia Allen. New York: John Wiley & Sons.
——. 1993b. "The Political Economy of Food: A Global Crisis." *New Left Review* 196: 29–57.
George, Susan. 1988. *A Fate Worse Than Debt.* London: Penguin.
Gilbert, Jess, and Raymond Akor. 1988. "Increasing Structural Divergence in U.S. Dairying: California and Wisconsin since 1950." *Rural Sociology* 53(1): 56–72.
Gomez, Sergio, and Walter Goldfrank. 1990. "World Market and Agrarian Transformation: The Case of Neo-Liberal Chile." Paper presented at the Twelfth World Congress of Sociology, Madrid, July.
Goodman, David, Bernardo Sorj, and John Wilkinson. 1987. *From Farming to Biotechnology.* Oxford: Basil Blackwell.
Hoogvelt, Ankie. 1982. *The Third World in Global Development.* London: Macmillan.
Kenney, Martin, Linda M. Lobao, James Curry, and W. Richard Goe. 1991. "Agriculture in U.S. Fordism: The Integration of the Productive Consumer." In *Towards a New Political Economy of Agriculture,* ed. William Friedland, Lawrence Busch, Frederick Buttel, and Alan Rudy. Boulder, Colo.: Westview Press.
Kloppenberg, Jack. 1984. "The Social Impacts of Biogenetic Technology in Agriculture: Past and Future." In *The Social Consequences and Challenges of New Agricultural Technologies,* ed. Gigi M. Gerardi and Charles C. Geisler. Boulder, Colo.: Westview Press.
——. 1988. *First the Seed: The Political Economy of Plant Biotechnology, 1492–2000.* Cambridge: Cambridge University Press.
Kneen, Brewster. 1990. *Trading Up: How Cargill, the World's Largest Grain Company, Is Changing Canadian Agriculture.* Toronto: N.C. Press.
LeClerq, Vincent. 1989. "Aims and Constraints of the Brazilian Agro-Industrial Strategy: The Case of Soya." In *The International Farm Crisis,* ed. David Goodman and Michael Redclift, pp. 275–91. London: Macmillan.
Mahler, Vincent A. 1986. "Controlling International Commodity Prices and Supplies: The Evolution of the United States Sugar Policy." In *Food, the State, and*

International Political Economy, ed. F. L. Tullis and W. L. Hollist, pp. 149–79. Lincoln: University of Nebraska Press.

Miner, William E., and Dale E. Hathaway. 1988. "World Agriculture in Crisis: Reforming Government Policies." In *World Agricultural Trade: Building a Consensus,* ed. Miner and Hathaway. Halifax, Nova Scotia: Institute for Research on Public Policy.

Morgan, Dan. 1979. *Merchants of Grain.* New York: Viking.

——. 1989. "Can Business Save Argentina?" *Business Week* 18 (September): 46–48.

Murphy, Tom. 1987. "Potato Capitalism: McCain and Industrial Farming in New Brunswick." In *People, Resources, and Power: Critical Perspectives on Underdevelopment and Primary Industries in the Atlantic Region,* ed. Gary Burrill and Ian McKay, pp. 19–29. New Brunswick, N.J.: Acadiensis Press.

Raikes, Philip. 1989. *Modernising Hunger, Famine, Food Surplus, and Farm Policy in the EEC and Africa.* London: James Curry/CIIR.

Rowe, J. W. F. 1965. *Primary Commodities in International Trade.* Cambridge: Cambridge University Press.

Salaman, Redcliffe. 1949. *The History and Social Influence of the Potato.* Cambridge: Cambridge University Press.

Sanderson, Steven, 1986. "The Emergence of the 'World Steer': Internationalization and Foreign Domination in Latin American Cattle Production." In *Food, the State, and International Political Economy,* ed. F. L. Tullis and W. L. Hollist, pp. 123–48. Lincoln: University of Nebraska Press.

——. 1989. "Mexican Agricultural Policy in the Shadow of the U.S. Farm Crisis." In *The International Farm Crisis,* ed. David Goodman and Michael Redclift, pp. 205–33. Oxford: Basil Blackwell.

Skinner, J. K. 1985. "'Big Mac' and the Tropical Rain Forests." *Monthly Review* 37: 25–32.

Sterngold, James. 1987. "Shaking Billions from the Beatrice Tree." *New York Times* September 6: sec. F, p. 1.

Strange, Marty. 1988. *Family Farming: A New Economic Vision.* San Francisco: Institute for Food and Development Policy.

Thomas, Robert J. 1985. *Citizenship, Gender, and Work: Social Organization of Industrial Agriculture.* Berkeley: University of California Press.

United States Department of Agriculture. 1986. *Outlook '87 Charts.* Washington, D.C.: Economic Research Service.

Vail, David. 1987. "Unique and Common Aspects of Sweden's Current Agricultural Crisis." Paper presented at Rural Sociological Society, Madison, Wisc., August.

van der Pijl, Kees. 1984. *The Making of an Atlantic Ruling Class.* London: Verso.

Wadekin, Karl-Eugene. 1990. *Communist Agriculture: Farming in the Far East and Cuba.* London: Routledge.

Woodhouse, Philip, and Ibrahima Ndiaye. 1990. "Structural Adjustment and Irrigated Food Farming in Africa: The 'Disengagement' of the State in the Senegal River Valley." Working Paper no. 20, Development Policy and Practice Research Group, Faculty of Technology, Open University, Milton Keynes, England.

Yotopoulos, Pan A. 1985. "Middle Income Classes and Food Crises: The New Food-Feed Competition." *Economic Development and Cultural Change* 33(2): 463–83.

12

Global Restructuring: Some Lines of Inquiry

Philip McMichael

The term *globalization* has gained wide currency in recent years. It generally refers to the worldwide integration of economic process and of space. Accordingly, globalization is perceived as a shift of power from communities and nation-states to international institutions such as transnational corporations and multilateral agencies, or even to supranational regional political organizations, such as the EC and NAFTA. Most significantly, it embodies images of growing homogeneity, or at least the diffusion of a singular market culture. In this sense, the discourse of globalization resembles the universal claims for the salience of the "self-regulating market" made by nineteenth-century liberal social theory (see Polanyi 1957). Its revival as conservative orthodoxy, legitimating the erosion of social entitlements (including wages) and managed economy, coincides with the growing power of unregulated global financial capital and political and economic restructuring (Bienefeld 1989).

In these various formulations, globalization is commonly depicted (and/or advocated) as a linear economic or political trend. Because of these shared conceptions of the meaning of globalization, it is not surprising to find commentators disputing such trends, or assumptions, with counterfactual arguments about the growing role of the state (Evans 1985), a new protectionist mood reducing the volume of international trade (Gordon 1988), and the subdivision of global economy into three competing trade blocs (Thurow 1992). Even so, the underlying conception of globalization remains uncontested.

This debate reveals the genuine confusion surrounding the phenomenon of globalization. This confusion is not lessened by the realization that we are experiencing a transitional phase in the history of world capitalism. Contradictory tendencies are therefore heightened, as capital accumulation processes undergo massive spatial and sectoral shifts accompanied by continual, unresolved political struggles around questions of political regulation, social distribution, and economic recovery. But this very conflict suggests some methodological procedures in grasping what underlying movements these manifest trends actually express. Such procedures might distinguish *residual* and *emergent* political-economic relations, on the one hand, and, on the other hand, identify combinations of these relations into new syntheses (see, e.g., McMichael 1992, 1993b).

From this perspective, it is useful to conceptualize globalization as a formative, and therefore contradictory, process. This concept avoids treating globalization phenomenally, as a set of linear (and contestable) trends. It also approaches it as the expression of a particular historical conjuncture where, as it were, the old is still dying and the new is yet to be born. In addition, such a conception allows us to specify this particular form of globalization vis-à-vis prior forms in preceding centuries (see, e.g., Wallerstein 1974, Wolf 1982). In short, globalization is more fruitfully understood not simply as a patterned, or restructured, outcome (e.g., the New International Division of Labor, informalization of economies, feminization of global labor forces, deindustrialization), but as a mechanism of restructuring also. That is, it is the key dynamic, at this particular time, reorganizing social, economic, political, and geopolitical institutions. Arguably, globalization is no longer simply one dimension of capitalist political economy, it now governs it (see, e.g., Sklair 1991). And it is not so much an inevitable development as symptomatic of the decline of a particular world order. The proof, as it were, lies in the pudding. This book is one such pudding: together, these essays exemplify this conception of the global transition.

Global Connections

As I suggested in the Introduction, these essays are framed explicitly or implicitly by a common temporal framework of restructuring, related to the collapse of the Bretton Woods system of national/international regulation. In this collapse lie at least three significant movements: shifts in regulatory conditions, new strategies of capital accumulation, and renegotiations of social and economic policy across the state system. These movements shape, and respond to, the new competitive conditions arising from the emergence of a global capital market, matched by new communications and transport technologies generating global production complexes and markets. In the presence of these new global conditions, all manner of local and regional responses occur.

The essence of global restructuring is precisely this differentiation. That is, if we perceive globalization as the subjection of historically uneven places to internationally competitive forces, where national regulatory systems have eroded, creating social and economic disjunctures (e.g., Appadurai 1990), then it will manifest precisely in the variety of responses. More fundamentally, this unevenness is a pivot for the pursuit of strategies of global competitiveness on the part of agencies such as firms or states. One salient example is the redivisioning of labor forces between "a relatively secure and protected minority, encompassed as a rule by enterprise corporatist relations, and a fragmented and relatively unprotected majority of nonestablished workers" (Cox 1987:281). This differentiation of labor tasks and employment conditions, termed a "process of Japanization" (Cox 1987:323), is a global phenomenon, crystallizing in combinations of high-tech and cheap labor, "as enterprises seek greater flexibility in adjusting production to . . . differentiated demand" (Cox 1987:321).

A striking example is the analysis by Kathleen Stanley in Chapter 5, detailing the reconstitution of meatpacking as a mass-production industry relying increasingly on flexible, unskilled, and mobile labor forces from another economic culture. As Stanley suggests, this reconstitution is the result of a competitive strategy, wherein new forms of centralization of meatpacking capital depend on imported labor as a wage-cutting device. The reformulation of the commodity (deboned/boxed beef) responds to the new segmented world market in a recomposition of the politicotechnical relations of production. In sum, a new global commodity requires a globally competitive labor force, with immigrant labor complementing (and replacing) local, organized labor. Methodologically, global restructuring is represented in the strategy of division and recombination of an uneven labor force in order to compete in a reconstituted world market. That is, it is a local (agro-industrial) response, and contribution, to changing global conditions.

It would be possible to treat each essay in these terms—for example, in Chaper 10, Pascal Byé and Maria Fonte's analysis of the range of possibilities in the restructuring of agricultural technologies conceptualizes the encounter between Fordist and post-Fordist forces in these terms. They present this encounter as an indeterminate juxtaposition of mechanical and biotechnical relations that is necessarily shaped by different political and cultural contexts. Perhaps more germane to a summary chapter, however, is some kind of synthesis of the story line beyond the presentation of thematic issues attempted in the Introduction. Harriet Friedmann's chapter (11) is a useful vehicle for situating these various connections methodologically, because it frames much of the content of the essays in this book: from the discussion of the politically managed metropolitan wheat complex (with differential implications for Swedish, East Asian, and Mexican agriculture), through identification of the origins of nontraditional agro-exports (such as fruits and vegetables) in metropolitan import substitutionism via the durable foods complex, to the internationalization of the livestock sector—both within the United States and elsewhere, such as Australia.

As suggested, one method of specifying global restructuring is to situate it precisely in historical time. The Introduction argued that the ending of the era of national regulation was a significant marker for understanding the global transition, and perforce for each essay. So we have two eras: the national and the postnational. We have a pretty good idea of what the former involved, but a quite incomplete idea and set of theories concerning the direction of the latter. But, I have argued, the juxtaposition of the two is the most promising place to begin, if only because it grounds our understanding of global restructuring as a contested and fluid process. Friedmann's essay exemplifies such a juxtaposition.

Friedmann's three complexes represent three forms of political economy that individually, and together, embody the contradictory relations within the world food economy. The first complex, wheat, represents the form of political-economy associated with the era of *national* regulation. Postwar protection of metropolitan farming sectors centered on the wheat complex, out of which came the massive surpluses that subsidized the new industrial labor forces underpinning Third World nation building. Although that complex certainly fostered Third World food dependency, it also became the counterpart national model for green revolutions within the Third World, especially in Mexico, India, Pakistan, Turkey, and Argentina, which have switched 84 percent of their total area planted in wheat to semidwarf seed varieties, displacing rain-fed crops such as coarse grains, oilseeds, and pulses, and dramatically reducing food imports (Vocke 1987).

The durable food complex was a by-product of the national model of subsidized farming, with its roots in *agro-industrialization*. Though the two processes of "appropriation" and "substitution" (Goodman, Sorj, and Wilkinson 1987) represent technological trends, these processes have been realized largely by means of political relations: farm protectionism for appropriationist tendencies, such as the application of new energy and capital inputs in industrial agriculture, and the international division of labor for substitutionist tendencies, such as the replacement of tropical products by scientifically extracted components of temperate crops. Not only did agro-industrialization amplify the consequences of the wheat complex, but also it accelerated a long-term erosion of the colonial division of labor, perhaps beginning with metropolitan sugar beet production (Mintz 1986:195, Tomich 1990) and expressed in our era in declining terms of trade for traditional tropical products (but see Talbot 1994).

The livestock complex (which would include poultry and pork), as the third of Friedmann's complexes, represents the *global* movement. At its inception in the postwar food regime as a national agricultural subsector, it already included global sourcing with feed crops. This national/international model spread as middle class diets developed within Third World countries, expanding the feed supply industry as a vehicle for differentiation among Third World states. In this context, with the application of international production and sanitary standards and global marketing, animal protein production has been restructured across the world into a global agro-industry.

The three complexes express three particular movements: the national movement (governed by the use-value of food), the global movement (governed by exchange-value), and the agro-industrial movement, which emerged in the former as a condition of the latter. In turn each movement is embedded within each complex. Their mutual dynamics constitute the world food economy as a formative and contradictory entity. More to the point, the global movement, as represented in the livestock complex, is currently restructuring the wheat and durable food complexes. Whatever linear trend there might have been (from national, through agro-industrial, to global) is undergoing a thorough reorganization, perhaps reversal, as national and local crop and livestock use-value) systems are subordinated to the global requirements of the animal protein (exchange-value) industry. The reorientation of staple foods as export crops for the global feedstuffs industry subordinates, indeed abstracts, place, or locale, in the service of globalized consumer markets. In short, the livestock complex, now recombining with the other two, exemplifies the extent, and social and political consequences, of commodification of food on a world scale.

But commodification, though perhaps a logical trend in capitalism is governed by political and geopolitical forces. It may infuse those forces, but they in turn constitute its limits and possibilities. This is the point of departure for suggesting some lines of inquiry that are opened by the collective contributions of this book. These lines of inquiry are (1) the question of regulation, (2) the specificity of agriculture, (3) food security issues, and (4) emerging geopolitical dynamics.

Some Lines of Inquiry

The Question of Regulation

The unifying theme, the era of postnational regulation, poses the question: what new forms of regulation of agro-food systems are likely to emerge? Most prominent is the free trade regime outlined in the 1992 Dunkel proposals to the GATT negotiations. The essence of these proposals is that trade liberalization is more than a multilateral implementation of the free movement of goods, rather it should include free trade–based investments, trade in services, and the rights of intellectual property, regulated by a new suprastatal organ (termed the World Trade Organization) to supervise these new areas and enforce compliance by member states. There are two striking issues regarding GATT. First, its long-term paralysis by national, or regional (CAP), resistances to general deregulation of agriculture, expressed in revolts by metropolitan farm groups from France to Japan. Arguably, such paralysis expresses the institutional residues of the national period that subvert the possibility of installing genuine global rules (McMichael 1993b). Second, the proposals to deregulate agriculture posit a form of posthegemonic global regulation, via multilateral institutions, of a world market organized by the transnationals (McMichael and Myhre 1991, Watkins 1991). In other words, GATT contention—likely to remain for some time—reveals the more profound difficulties facing the world in transition.

Within the neoliberal climate surrounding the Uruguay Round, it has been virtually impossible to inject an alternative conception of global regulation, such as that set forth by the World Food Board proposal in 1947 (see Chapter 11, Raghaven 1990). Nevertheless, recent attempts to rehabilitate Keynesianism, forced upon metropolitan states by the exigencies of global competition, may well drift upward, so to speak, to frame a managed global economy. Such supranational regulation might replace declining national forms of regulation simply because it can crystallize the collective relations among states in a world economy organized increasingly by transnational companies whose global strategies now supersede the managed trade of the national regulation era.

The issue is whether and to what extent states would surrender economic sovereignty to supranational institutions. Some argue that the surrender has begun and in fact defines the current transition (e.g., Cox 1987, Kolko 1988, Picciotto 1990, Cerny 1991, McMichael and Myhre 1991), and further, that this surrender involves an erosion of democratic gains, facilitating this transition (Held 1991, Gill 1992). Alternatively, this movement is perceived as multifaceted, transcending "the old Westphalian concept of a system of sovereign states;" in this movement, sovereignty "has gained meaning as an affirmation of cultural identity and lost meaning as power of the economy" (Cox 1992:34). This shift allows new possibilities for social movements that extend, or transcend, the traditional (nation-centered) class movements of the previous century (Arrighi 1990, Wallerstein 1992). In this context, we may see the emergence of new claims being made on national states for protections against market forces, such as those addressed in David Vail's Chapter 2 and David Myhre's Chapter 6, under the new constraints of balancing global economic viability and domestic legitimacy. But this negotiation may well occur in a new terrain of politics, where supranational regulation may either weaken or empower locally based political movements.

Alternatively, or in some complementary relationship to managed global economy, regional configurations—whether at the macroregional or microregional level (see Cox 1992)—are already forming. These configurations concern regional economies, such as the NAFTA, EC, MERCUSOR (the common market of the Southern Core of Latin America) or Association of Southeast Asian Nations (ASEAN), regulating currency movements and trade imbalances among contiguous states at a higher level of resolution than the national economy, with some new concept of equity and social justice that is multinational. Unsurprisingly, the threat of such regionalism spurs transnational corporate investment. Microregional movements, such as the Toronto Food Council and the Northern Italy agro-food complex (Friedmann 1993), suggest the possibility of linking planned land use with healthy diets and local food processing to reassert the advocacy of local (food) self-reliance associated with the politics of Food First (Lappe and Collins 1978).

These various alternatives to national regulation and their differential implications for power and organization of agro-food systems are key areas of contention and therefore of inquiry. At the same time, considerable unilateral reform of metropolitan farming (including land set-aside, decoupling, subsidy reduction) has taken place, despite the Uruguay Round and its institutional shortcomings (Petit 1985:70, Paarlberg 1992:42). These reforms reflect both a general fiscal crisis and changing power/discourse balances within particular states—such as Sweden's food policy reforms documented in Chapter 3 by David Vail and the Dutch reversal of centuries of land reclamation in order to limit environmental damage in a context of new land set-aside policies (Simons 1993). Reform of agricultural policy may stem from the growing recognition of the fiscal, environmental, or global unsustainability of subsidized metropolitan agriculture, but the actual direction of reform will depend ultimately on the relative power of the relevant constituencies. Further, deregulation is double-edged insofar as it may pare public budgets, taxes, and consumer prices on the one hand, but on the other centralize control of global food systems in the hands of global corporations.

It is clear that there must be new forms of regulation of the world food economy, but it is unclear what these will be and what directions they will take. Three possible scales of analysis must be considered in order to track this process: global, regional, and local. At the global scale, examination of the course of the GATT negotiations is necessary, as the new rules of international agro-food trade stem from these talks. Two related areas of inquiry are the role of large-scale shifts in the structure of world agriculture and the negotiation between states and transnational corporate interests. The emergence of new agro-exporters, and new import zones such as East Asia and Eastern Europe, are reformulating agricultural power relations in the world food economy, and these will surely affect transnational corporate strategies. These issues have salience at the regional scale, where inquiry might well consider to what extent, and how, regional developments anticipate, or retard, elaboration of global rules. Finally, the local scale is the likely arena for response to these regional and global movements, and it will be useful to compare different farm and political systems in understanding how national and subnational social movements obstruct and/or complement supranational regulations.

The Specificity of Agriculture

It has become commonplace to argue that the sectoral boundaries of agriculture are blurred by agro-industrialization (e.g., Goodman, Sorj, and Wilkinson 1987, Kloppenburg 1988, Friedmann and McMichael

1989; but see Baxter and Mann 1992). Related to this claim is the argument that the classical agrarian question animating political debates at the turn of this century has yielded to questions concerning the politics of the global division of agricultural labor (McMichael and Buttel 1990), and the food and green questions (Goodman and Redclift 1991, Bonnano 1992, Whatmore, Munton, and Marsden 1991, Buttel 1992). From a historical perspective, this movement is evident within the debates concerning the CAP, where initial productivism and food self-sufficiency have given way to "a more socially oriented political agenda" (Bonnano 1991:557). In general, where agro-food systems disengage from their rural origins or contexts, new questions concerning the social role of rurality and the sustainability of the natural environment emerge.

The question of agriculture's specificity also concerns its historic spatial role. There is something elemental about agriculture—not just in its natural biological limits to capital, but also in its association with place, culture, and sustenance—that elevates it to the foreground in the politics of this transition. Modern nation-states, whether for reasons of territorial integrity or development strategy, have institutionalized a stake in agriculture that often exceeds that of other economic sectors. That stake is under question as economic sovereignty and development strategies undergo reconsideration in this global transition. Fueling this debate are the GATT negotiations and widespread resistances to agricultural liberalization. Not only do protected political constituencies resist disenfranchisement, but also liberalization is perceived as a cultural threat. Arguably, bringing agriculture under a GATT regime is profoundly symbolic of the move to legitimize world market integration because of agriculture's identification with place and nation. Place still governs the bulk trade linked to the grain and meat complexes of national farm sectors. Alternatively, the new agro-exports, such as specialty foods from the south, represent transnational trading patterns based on "intra-industry specialization, intra-firm investment decisions, and niche markets" (Josling 1992:4). GATT liberalization is likely to privilege the transnational movement against residual national economic organization. Ironically, this privileging is especially likely to occur where current metropolitan farm policy reforms uncouple farm income supports from commodity pricing—a trend actually favored by transnational food companies as it would depress production costs (Kneen 1990:96).

In this confrontation, questions emerge around the political dimensions of agriculture, where commodification clashes with social protectionism. As a modern example of social protectionism, the Greens movement is arguably self-consciously universalistic in identifying the preservation of farming and rural economy as an act of ecological sustainability. At the same time, the Greens movement has arisen in a historical context where social justice discourse has declined precipitously as a consequence of economic restructuring mechanisms that have "hollowed" out social entitlements and organized labor (Buttel 1992). Whether and to what extent environmental movements can anchor concerns for ecological sustainability in concerns for social justice remains to be seen.

The challenge to farm protectionism is a challenge to the social protectionism identified with the twentieth-century nation-state. As such, it also raises the puzzle of the specificity of agriculture, that is, the ways in which it is politically and socially constructed. Future research on the question of the survival of agricultural systems and rurality will have to examine this through the filter of the political organization of agriculture—on a national, national-comparative, and global scale. This political focus includes issues of regulation of farming, biotechnologies, and food quality. It is very likely that the struggle over the extent of commodification of the environment and of social life will be rooted in questions of control and revival of local food systems. In the immediate future, it will be important to examine this struggle in its different forms across the world—one possible subdivision being between its resonation in the north with issues of diversity of foods and natural environments, and in the south with issues of autonomy and stability of rural populations.

Food Security Issues

Food security has always been a political question (Spitz 1985). Although food supply politics centered on the peasant community as the repository of moral economy and defender of the right to food during the transition to capitalism (Tilly 1975:389), more recently food riots have characterized Third World urban politics in response to the depredations associated with the structural adjustment policies of the 1980s (Walton 1990). Arguably, this modern social movement arises from the combination of two historical trends, one secular and the other cyclical/transitional.

Briefly, the secular trend is that marking the threshold of modernity, where early modern state formation processes first began to increase claims on peasant surpluses to guarantee food to noncultivators located in emerging towns. The comprehensive redistributive mechanisms of precapitalist societies were regulated by conceptions of moral economy, where rights to subsistence were customary, or embedded in the structure of the community (Polanyi 1957, Spitz 1985). The political distancing of states and urban consumers from producer rights over time and the growing pressures on states to expand military and bureaucratic organization intensified food requisitioning and the undermining of moral economy (Wolf 1966, Tilly 1975). Discriminatory price policies, obscuring urban extraction from the countryside, and the growing commercialization of agriculture have not only institutionalized "urban bias" (Lipton 1977) but also reconstituted the peasantry (and its moral claims) as modern producer, but second-class citizen (cf. Sorj and Wilkinson 1990). That secular process has of course been compressed into several decades for much of the Third World. The period from 1950 to 1975 "saw the most spectacular, rapid, far-reaching, profound, and worldwide social change in global history. . . . [This] is the first period in which the peasantry became a minority, not merely in industrial developed countries, in several of which it had remained very strong, but even in the Third World countries" (Hobsbawm 1992:56).

Superimposed on this secular modernization-style social trauma visited on the Third World peasant in the postwar era were the draconian policies of structural adjustment of the 1980s. These mandated reductions in government supports for peasant producers opened economies vis-à-vis imports of artificially cheapened food and encouraged new agro-exports to retire debt. Each of these measures has intensified the undermining of conceptions, practices, and the resources of domestic food security.

New social movements have emerged in the countryside: whether positive mobilizations to reclaim or buttress social entitlements (as depicted in Chapter 6) and local sustainable agrarian practices (Rau 1991) or demobilization and withdrawal of rural communities as documented by Fantu Cheru (1989). The Mexican case may be culturally specific, given the legacies of the Mexican revolution in the political system—although these are being rapidly erased through the current dismantling of the *ejido* system (Barkin 1993). The demobilization process in turn may be specific to Africa, given the legitimacy crisis of stark discrepancies between development promises by multilateral agencies and African governments alike and the reality of severely declining agricultural production and living standards in the sub-Saharan sections of that continent (Raikes 1988, Bradley and Carter 1989). Nevertheless, in the wake of industrial fetishism and agro-industrialization during the postwar state-building process, there are promising new movements to revitalize rural economy through diversification (Koppel, Hawkins, and James, 1994) that may in turn revitalize and anchor local food systems in the future.

At the same time, food riot movements have become commonplace in Third World cities (Walton 1990) as rapidly expanding urban populations have outstripped urban food systems. Adding to the food deficits have been reorganizations of urban food systems by retailers and food processors, often rendering urban food systems increasingly inaccessible to the urban poor via such commodification (Drakakis-Smith 1991). Ironically, the politics of urban stability and food supply/distribution problems have reinforced Third World food dependency as cheap imported foods resolve legitimacy problems in the short run. In addition, commodification of food systems can be a conduit for agribusiness contracts with international grain traders (see, e.g., Andrae and Beckman 1985).

Unlike manufactured goods, and some services, for which comparative advantage may be less problematic, food accessibility is a powerful symbol of political legitimacy and sovereignty. That is, the use value of food potentially outweighs its exchange value in the cultural significance of all commodities. Thus food security movements at the national and subnational level are likely to remain a significant source of resistance to global conceptions of food security, the most virulent form of which is the proposal by neoclassical adherents to deregulate world markets and subordinate food to exchange value.

In short, the issue of food security will likely center on the growing tension between food entitlement and food commodification movements, confirming the fundamentally political character of food. An appropriate area of inquiry is examining the impact of food riots and food security movements on government policies, especially what priorities emerge between cheap food imports and the renewal of local agricultures. In this sense, the so-called food question is likely to vary across nation-states, but will routinely implicate government policies that mediate agribusiness interests and domestic farm populations. Research that uncovers local solutions, and the political-economic conditions for these, will have important demonstration effects, not the least being correctives to the neoclassical project, which reduces food policy questions to economic discourse. In the larger sense, the politics of the food question could arguably become the central issue for the twenty-first century—not only because it concerns political entitlement in both the formal (sustenance) and substantive (autonomy) senses, but also because it implicates sustainable environmental practices. It will therefore be necessary to examine the reshaping of rules for the world economy as a broad, millennial question of human and food security politics.

Emerging Geopolitical Dynamics

The future of agricultural producers in general, and of specialized producers such as grain producers in particular, is likely to depend upon geopolitical alignments. Certainly neoclassical economic theorists argue that an open world food system, such as that envisaged in the GATT Uruguay Round, allows the reassertion of comparative advantage. This is not simply a question of place differentials, as technical and institutional innovations have lowered transport costs and internationalized information, thereby enabling consumers to "tap supplies of food and agricultural products in almost any part of the world" (Schuh 1987:73). From this perspective, northern agricultural deregulation and southern currency devaluation would combine to raise world market prices (Alexandratos 1988:15) and expand world trade, especially southern agro-exports (Schuh 1987:83). The unstated premise is that much of this trade is managed by the transnationals. But this perspective needs tempering with the changing geopolitics of agro-food demand and supply and attention to the composition of world trade.

In the likely tidal change in the organization of the world food economy there are several possibilities and emerging trends. These include first, the reorganizations in the former Soviet Union and China and their likely impact on trade; second, the fate of the surplus agro-exporters such as the United States and the EC in the world market; third, the related prospects of southern agro-exporters; fourth, the growing significance of East Asian, and especially Japanese, agro-food importing, which replicates to some extent Britain's role in the nineteenth-century world economy; and fifth, the possibility of regional agro-food integration underwriting trade bloc patterns on the one hand, and transnational global/regional sourcing on the other—in particular redirecting corporate investment into food processing operations in the south as northern protectionism declines.

Of course it is difficult to predict future outcomes, but we can perhaps identify what must be examined in order to grasp ongoing trends. There are three interrelated processes that need attention if one is to trace emerging configurations in the world food economy. One process is geopolitical and concerns shifts in production and trade as well as the rules governing world agro-food markets. Another is technological, and concerns questions such as the extent to which appropriation of natural aspects of agriculture and substitution of agricultural products proceed, affecting the location and destination of agricultural commodity production. The third process is economic, and concerns accumulation and market strategies of food companies.

Potentially, both Russia and China are significant grain exporters, of wheat and corn respectively, and importers of livestock feedstuffs. China, in particular, demonstrates a comprehensive process of decollectivization as agricultural management has shifted to the household level (Selden 1993). The consequences of this have been rising agricultural productivity, especially in the early 1980s, and a diversification of production and diets away from grain toward animal protein and horticultural products. General economic privatization, and the broadening of the concept of special economic zones (Pepper 1988), has fueled expansion of commercial and industrial centers in the coastal provinces, supported by a population newly mobile both geographically and socially. Unsurprisingly, meat consumption per capita more than doubled through the 1980s. Since animal protein accounts for only 15 percent of dietary protein in China, the livestock industry is a likely source of growth, suggesting that China could become an increasingly consequential importer of feedstuffs, including corn, since meat prices considerably exceed feedstuff prices. By the same token, it is possible that this dietary shift would encourage the Chinese to replace some rice grain imports (USDA 1992b:38, 45). This, and the possibility of trade liberalization, predicted to raise grain prices in the short term (UNCTAD 1990:47), would encourage the expansion of Chinese grain production. Given the size of China, this development would have considerable impact on the world food economy. In particular, China has already become a consequential supplier of a range of foodstuffs (some processed) to the Japanese market, which is a new growth pole in the world food economy.

In the former Soviet Union, under the severe inflationary conditions of liberalization and the reduction of farm subsidies, agriculture has stagnated. Decollectivization in Russia, though proceeding apace, accounted for only 2 percent of the 211 million hectares of agricultural land—and many of the new private farmers were urban residents (USDA 1992a:3, 14–15, 17). And, given that two generations of Russian farmers have no experience of private farming and face the resistance of regional state farm officials to the surrender of their administrative power (Selden 1993), the return of Russia to its historic grain export role will depend on the efficacy of the reforms. These include privatization of the huge grain-purchasing agency Roskhleboprodukt, recipient of the lowest farm credits, and the end of farm subsidies (Erlanger 1993:9).

In the meantime, the region encompassed by the former Soviet Union is likely to continue to be a significant grain importer—but, in the short run, only to the extent that it continues to receive credits from the G-7 countries. In 1992, for example, the United States supplied one-third of the wheat and one-half of the coarse grains imports to the former Soviet Union), and the EC supplied one-fifth each of the wheat and coarse grain imports (USDA 1992a:30, 59). There is a clear global political objective in U.S. exports to the region in support of the reforms; however, a possible resolution to the GATT dispute may be a medium-term arrangement whereby EC surpluses are sold to the region "cushioning the shock to EC farmers of a more substantial cut in agricultural export subsidies." Former EC markets would thereby open up to Cairns Group agro-exporters (Bhagwati 1991:116–17), complemented by Eastern European states trading grain for Russian oil. Because of the unpredictable outcome of privatization of land in Russia, we may expect to see more transnational corporate investment in grain production elsewhere in middle-income Southern states like Mexico, Brazil, and Argentina.

With respect to the question of the future of grain markets, it is not yet clear whether and to what extent metropolitan farm support systems will be eliminated, although the likelihood of a GATT-based market regime is presaged in the erosion of national regulation (McMichael 1992, 1993b). Implementation of such a regime may take considerable time. Though farm supports continue, the current trend of expensive, managed dumping will also continue to exacerbate southern food dependency (Watkins 1991), the anarchy of world grain markets (possibly offset by Russian importing), and budget deficits around the world. Perhaps in the end the financial ramifications of these trends will force solutions more rapidly than bilateral conflicts over farm policies.

The free trade thrust, to include agriculture in the GATT, was initiated by the United States. It is curiously reminiscent of Britain's free trade policy in the mid-nineteenth century, when Britain attempted to elaborate a colonial system writ large, based in its own central role as "workshop of the world" in exchange for raw material and agricultural imports from all colonial (and ex-colonial) regions (McMichael 1984). In this case, U.S. initiation of the Uruguay Round, in part to recapture shrinking world agricultural commodity markets, is evidently based in a "breadbasket of the world" conception (McMichael 1994). It is generally believed that trade liberalization would favor exports of the United

States and members of the Cairns Group, to the disadvantage of EC and Japanese trade balances (Rayner, Ingersent, and Hine 1990:7). U.S. rivalry with the EC in world food markets is legendary, and ideological in the sense that the EC is identified as "benefiting from 'unfair' advantages" stemming from export subsidies (Petit 1985:58). The U.S. neo-mercantilist deployment of the Super 301 clause of the Omnibus Trade Act of 1988, claiming right of retaliation against any perceived unfair trade practices (Winters 1990:1301), in addition to the 1985 Export Enhancement Program, fueled the intense trade fiction of the 1980s.

In fact, a free trade regime may be a double-edged sword for the Unites States. Transnational food companies, active supporters of liberalization to reduce commodity (input) prices, would be in a position to intensify competition between U.S. and alternative producers in world markets. Intensified price, or product, competition is possible in this scenario. In the 1980s, for example, Brazil and Argentina, with alternative oilseeds such as rapeseed, cotton seed, and peanuts, captured 50 percent of world market share from U.S. soybean exporters (Rosson et al. 1989, Crowder 1990). In addition, agro-exporters such as the United States may find their surpluses have become liabilities (Friedmann 1993). Evidence indicates the possibility that China, Eastern Europe, and Egypt use their import power to obtain the most favorable terms of trade among competing regions coordinated by grain companies (International Wheat Council 1988, Report 16; Pick and Park 1991).

In the area of animal feeds, such as secondary cereals and oil products, the United States competes favorably with the EC (Lambert 1990). Again, new biotechnologies in combination with the growing global sourcing of intensive animal protein production have intensified competition from Brazil, China, and Thailand as rivals to traditional cereal exporters like the United States. Not only have feed crops been replacing food crops globally (Barkin, Batt, and DeWalt 1990), but also grain substitutes like cassava, corn gluten, and citrus pellets have become important export commodities (Hathaway 1987:30). As Friedmann suggests in Chapter 11, these crops fuel the global competitive strategies of food and chemical companies, the latter requiring plant-derived feedstocks as substitutes for fossil hydrocarbon feedstocks in producing chemicals and fuels (Buttel 1990:176–77).

The U.S. "breadbasket" strategy, although somewhat anachronistic in a world food economy where the center of gravity is shifting away from bulk, low-value products toward a brisk trade in high-value products, continues to be based on the expectation of rising middle class demand in less developed countries for animal protein products. The assumption is of a rising demand for feedstuffs, given the technical limits to local cereal production in rain-fed regions in the Third World. Certainly imports of feed grains exceeded food grains in the 1980s, especially in the middle-income states such as South Korea, Singapore, Indonesia, Mexico, and Brazil (de Janvry, Sadoulet, and White 1989). U.S. data suggest that the United States perceives American cereal farmers to have a comparative advantage (Vollrath and Scott 1991), in addition to a general faith in the universality of the dietary implications of the U.S. agro-food model.

What we do not know is for how long this agro-food model is sustainable, given the worldwide strain on resources and dietary inequalities associated with the animal protein industry (Durning and Brough 1992), and whether affluent consumption can be sustained alongside the deterioration of natural resources and the conditions of life of disenfranchised and dispossessed peoples that accelerated in the 1980s (Durning 1990, Brown 1993).

In addition, U.S. predictions about the complementarity of American agriculture and global dietary sifts are based perhaps in residual trends associated with managed farm surpluses, a model increasingly challenged by transnational firms and the GATT negotiations. The emergent trend, in which agro-industries are reconstructed around global corporate strategies rather than around national economic complexes, may well undermine the unprotected northern farmer in the event of a GATT free trade regime (see Buttel 1989). Such undermining is already evident in the reconstruction, by transnational firms, of animal protein production in the United States (see Chapter 5), in Thailand (McMichael 1993a), and Australia (see Chapter 3, Ufkes 1993)—the latter two regions successfully rivaling U.S. exports in the new specialty markets in East Asia. It is also prefigured in the rush of U.S. Japanese, and European firms to invest in food processing in Mexico. This movement is associated with reversals of constraints on foreign investment and the implementation of new land legislation undermining the ejido system and stimulating agro-export zones subcontracting with food processors. U.S. investment in Mexican food processing, after declining 17 percent annually through most of the 1980s, rebounded by 81 percent in 1989 (Carlsen 1991:21). This reversal expresses a new trend perhaps, in that during the 1980s food processors relocated investment from the south into a process of restructuring, and centralization, of corporate food capital in the metropolitan states (as discussed in Chapter 4), given "the decline of the macroeconomic environment in most developing countries" in that decade (Rama 1992:269). But Mexican deregulation perhaps symbolizes 1990s movements for regional integration, as stabilizing mechanisms in the process of global restructuring. These are undoubtedly two-sided, in the sense that regional integration is a defensive global strategy on the part of states such as the United States, Japan, and the EC, at the same time as it is a spur to foreign investment, and hence to further global economic organization through trade expansion and joint venture capitalism.

It is clear that there are various geopolitical movements, or initiatives, under way in the world food economy. It is impossible to predict outcomes with any degree of certainty. Nevertheless, in keeping with the methodological axiom of this book, a useful rule of thumb for future research is to juxtapose emergent trends with residual trends in order to sort out likely directions and points of conflict. One possible line of inquiry is that of juxtaposing the declining export regime centered on the United States with the emerging array of agro-export systems built on competitive (e.g., grains) and alternative (high value specialty foods) principles, or with the rising import complex centered on East Asia (see Friedmann 1993, McMichael 1993a). This juxtaposition frames questions about the geopolitical contours of any future food regime. Within, or alongside, that framework the various scenarios for Russia and Eastern Europe, China, and the EC-Africa, Japan-ASEAN, and NAFTA axes constitute important regional studies necessary to undergird the broader global questions. For example, to what extent will regional configurations shape regional agricultures, and how much influence will they have in qualifying global agricultural commodity flows? And these configurations in turn may implicate previous questions concerning new forms of national and supranational regulation which will affect food security issues and transnational corporation strategies alike.

Conclusions

The world is clearly at a significant threshold. Crossing that threshold is increasingly problematic, as there is a fundamental imbalance of power. Accordingly, there is perhaps an unwillingness to rethink the rationalist, consumption-oriented industrial model that has driven state-sponsored "development" strategies over the past century, has established such deep divisions among people along class, gender, and ethnic lines, and will undermine sustainability of life itself in the long term.

This book's focus on global restructuring directly addresses the turning point that we face today. It is not simply a kind of neoclassical inspired economic "spring cleaning" or a reassertion of the historical power of the north over the south. Nor is it simply a movement from, or a conjunction of, mass production to flexible lean production that escapes the standardization of the Fordist period. It is also a reformulation of political relations within and among states, via global integration. Such global relations are enforced through the market by transnational companies or multilateral agencies with new regulatory projects founded in abstract economic principles rather than the politically engaged social constituencies associated with the nation-state. How these relations and their countermovements among citizen, producer, worker, and consumer groups will evolve is an open question. But one thing is definite: questions of environmental sustainability and food security will be central, prompting a renewed focus on the social organization of agriculture and the delivery of its products. Agro-food systems will play a central role in the political change defining this era of transition.

References

Alexandratos, Nikos. 1988. *World Agriculture: Toward 2000*. New York: New York University Press.

Andrae, Gunilla, and Bjorn Beckman. 1985. *The Wheat Trap*. Chicago: Zed Press.

Appadurai, Arjun. 1990. "Disjuncture and Difference in the Global Political Economy." *Public Culture* 2(2): 1–24.

Arrighi, Giovanni. 1990. "Marxist Century, American Century: The Making and Remaking of the World Labour Movement." *New Left Review* 179: 29–64.

Barkin, David. 1993. "The New Shape of the Countryside: Agrarian Counter-Reform in Mexico." Unpublished manuscript.

Barkin, David, Rosemary Batt, and Billie DeWalt. 1990. *Food Crops vs. Feed Crops: The Global Substitution of Grains in Production*. Boulder, Colo.: Lynne Reiner.

Baxter, Vern, and Susan Mann. 1992. "The Survival and Revival of Non-Wage Labor in a Global Economy." *Sociologia Ruralis* 32(2/3): 231–41.

Bhagwati, Jagdish. 1991. "Jumpstarting GATT." *Foreign Policy* 83: 105–18.

Bienefeld, Manfred. 1989. "The Lessons of History and the Developing World." *Monthly Review* 3: 9–41.

Bonnano, Alessandro. 1991. "From an Agrarian to an Environmental, Food, and Natural Resource Base for Agricultural Policy: Some Reflections on the Case of the EC." *Rural Sociology* 56(4): 549–64.

——. 1992. "The Restructuring of the Agricultural Food System: Social and Economic Equity in the Reshaping of the Agrarian Question and the Food Question." *Agriculture and Human Values* 8(3): 72–82.

Bradley, Philip N., and S. E. Carter. 1989. "Food Production and Distribution—and Hunger." In *A World in Crisis? Geographical Perspectives*, ed. R. J. Johnston and P. J. Taylor, pp. 101–24. London: Basil Blackwell.

Brown, Lester R. 1993. "A New Era Unfolds." *State of the World 1993*, ed. Lester Brown, pp. 3–21. New York: Norton.

Burfisher, M. E., M. D. Missiaen, and A. L. Blackman. 1991. "Less Developed Countries' Performance in High-Value Agricultural Trade." Staff Report. Washington, D.C.: USDA-ERS.

Buttel, Frederick H. 1989. "The U.S. Farm Crisis and the Restructuring of American Agriculture: Domestic and International Dimensions." In *The International Farm Crisis*, ed. David Goodman and Michael Redclift, pp. 46–83. New York: St. Martin's Press.

——. 1990. "Biotechnology and Agricultural Development in the Third World." In *The Food Question: Profits or People?* ed. Henry Bernstein, Ben Crow, Maureen Mackintosh, and Charlotte Martin, pp. 163–80. New York: Monthly Review Press.

——. 1992. "Environmentalization: Origins, Processes, and Implications for Rural Social Change." *Rural Sociology* 57(1): 1–28.

Carlsen, L. 1991. "Reaping Winter's Harvest." *Business Mexico* 1(3): 20–23.

Cerny, Philip G. 1991. "The Limits of Deregulation: Transnational Interpenetration and Policy Change." *European Journal of Political Research* 19: 173–96.

Cheru, Fantu. 1989. *The Silent Revolution in Africa: Debt, Development, and Democracy*. Chicago: Zed Press.

Cox, Robert W. 1987. *Production, Power, and the World Order: Social Forces in the Making of History*. New York: Columbia University Press.

——. 1992. "Global Perestroika." In *Socialist Register*, ed. Ralph Miliband and Leo Panitch, pp. 26–43. London: Merlin Press.

Crowder, B. M. 1990. "Government Programs for Soybeans." *National Food Review* 13(1): 32–36.

de Janvry, Alain, Elizabeth Sadoulet, and T. K. White. 1989. *Foreign Aid's Effect on U.S. Farm Exports: Benefits or Penalties?* Washington, D.C.: USDA-ERS.

Drakakis-Smith, David. 1991. "Urban Food Distribution in Asia and Africa." *The Geographical Journal* 157(1): 51–61.

Durning, Alan B. 1990. "Ending Poverty." In *State of the World*, ed. Lester R. Brown, pp. 135–53. New York: Norton.

Durning, Alan T., and Holly B. Brough, 1992. "Reforming the Livestock Economy." In *State of the World 1992*, ed. Lester Brown, pp. 66–82. New York: Norton.

Erlanger, Steven. 1993. "Bread Prices Rise: Russians Resigned. Free-Market Economists Try to Relieve State of Burden of Broad Farm Subsidy." *New York Times*, October 17:A9.

Evans, Peter. 1985. "Transnational Linkages and the Economic Role of the State: An Analysis of Developing and Industrialized Nations in the Post–World War II Period." In *Bringing the State Back In*, ed. Peter Evans, Dietrich Rueschemeyer, and Theda Skocpol, pp. 192–226. New York: Cambridge University Press.

Friedmann, Harriet. 1993. "The Political Economy of Food: A Global Crisis." *New Left Review* 197: 29–57.

Friedmann, Harriet, and Philip McMichael. 1989. "Agriculture and the State System: The Rise and Decline of National Agricultures, 1870 to the Present." *Sociologia Ruralis* 29: 93–117.

Gill, Stephen. 1992. "Economic Globalization and the Internationalization of Authority: Limits and Contradictions." *Geoforum* 23(3): 269–83.

Goodman, David, and Michael Redclift. 1991. *Refashioning Nature: Food, Ecology and Culture.* Oxford: Basil Blackwell.

Goodman, David, Bernardo Sorj, and John Wilkinson. 1987. *From Farming to Biotechnology: A Theory of Agro-Industrial Development.* Oxford: Basil Blackwell.

Gordon, David. 1988. "The Global Economy: New Edifice or Crumbling Foundations?" *New Left Review* 168: 24–65.

Hathaway, Dale E. 1987. *Agriculture and the GATT: Rewriting the Rules.* Washington, D.C.: Institute for International Economics.

Held, David. 1991. "Democracy, the Nation-State, and the Global System." *Economy and Society* 20: 1–56.

Hobsbawm, Eric. 1992. "The Crisis of Today's Ideologies." *New Left Review* 192: 55–64.

International Wheat Council. 1988. "Wheat Support Policies and Export Practices in Five Major Exporting Countries." *International Wheat Council, Secretariat Paper* 16: 1 – 71.

Josling, Tim. 1992. "Emerging Agricultural Trade Relations in the Post–Uruguay Round World." Presentation to the Faculty of Political Science, University of Rome, June.

Kloppenburg, Jack, Jr. 1988. *First the Seed: The Political Economy of Plant Biotechnology.* Cambridge: Cambridge University Press.

Kneen, Brewster. 1990. *Trading Up: How Cargill, the World's Largest Grain Company, Is Changing Canadian Agriculture.* Toronto: N.C. Press.

Kolko, Joyce. 1988. *Restructuring the World Economy.* New York: Pantheon.

Koppel, Bruce, John Hawkins, and William E. James, ed. 1994. *Development or Deterioration?: Work in Rural Asia.* Boulder, Colo.: Westview Press.

Lambert, Frederick. 1990. "Is European Agriculture Competitive in the Face of Its North American Competitors? An Analysis Based on a Comparison of Producer Prices." *Economie-Rurale* 196: 23–31.

Lappe, Francis Moore, and Joseph Collins. 1978. *Food First: The Myth of Scarcity.* New York: Ballantine Books.

Lipton, Michael. 1977. *Why Poor People Stay Poor: Urban Bias in World Development.* London: Temple Smith.

McMichael, Philip. 1984. *Settlers and the Agrarian Question: Foundations of Capitalism in Colonial Australia.* Cambridge: Cambridge University Press.

——. 1992. "Tensions between National and International Control of the World Food Order: Contours of a New Food Regime." *Sociological Perspectives* 35(2): 343–65.

——. 1993a. "Agro-Food Restructuring in the Pacific Rim: A Comparative-International Perspective on Japan, South Korea, the United States, Australia, and Thailand." In *Pacific-Asia and the Future of the World-System,* ed. Ravi Palat, pp. 103–16. Westport, Conn.: Greenwood Press.

——. 1993b. "World Food System Restructuring under a GATT Regime." *Political Geography* 12(3): 198–214.

——. 1994. "GATT, Global Regulation and the Construction of a New Hegemonic Order." In *Regulating Agriculture,* ed. Philip Lowe, Terry Marsden, and Sarah Whatmore. London: David Fulton.

McMichael, Philip, and Frederick H. Buttel. 1990. "New Directions in the Political Economy of Agriculture." *Sociological Perspectives* 33: 89–109.

McMichael, Philip, and David Myhre. 1991. "Global Regulation vs. the Nation State: Agro-Food Systems and the New Politics of Capital." *Capital & Class* 43: 83–105.

McMichael, Tony. 1993. *Planetary Overload: Global Environmental Change and the Health of the Human Species.* Cambridge: Cambridge University Press.

Mintz, Sidney. 1986. *Sweetness and Power: Sugar in Modern History.* New York: Vintage.

Paarlberg, Robert L. 1992. "How Agriculture Blocked the Uruguay Round." *SAIS Review* 12(1): 27–42.

Pepper, Suzanne. 1988. "China's Special Economic Zones." *Bulletin of Concerned Asian Scholars* 20(3): 2–21.

Petit, Michel. 1985. *Determinants of Agricultural Policies in the United States and the European Community.* Washington, D.C.: International Food Policy Research Institute.

Picciotto, Sol. 1990. "The Internationalization of the State." *Review of Radical Political Economics* 22(1): 28–44.

Pick, D. H., and T. A. Park. 1991. "The Competitive Structure of U.S. Agricultural Exports." *American Journal of Agricultural Economics* 73(1): 133–41.

Polanyi, Karl. 1957. *The Great Transformation: The Political and Economic Origins of Our Times.* Boston: Beacon.

Raghavan, Chakravarthi. 1990. *Recolonization: GATT, the Uruguay-Round, and the Third World.* Penang, Malaysia: Third World Network.

Raikes, Philip. 1988. *Modernizing Hunger: Famine, Food Surplus, and Farm Policy in the EEC and Africa.* London: Catholic Institute for International Affairs.

Rama, Ruth. 1992. *Investing in Food.* Paris: OECD.

Rau, Bill. 1991. *From Feast to Famine: Official Cures and Grass Roots Remedies to Africa's Food Crisis.* Chicago: Zed Press.

Rayner, A. J., K. A. Ingersent, and R. C. Hine. 1990. "Agriculture in the Uruguay Round: Prospects for Long-term Trade Reform." *Oxford Agrarian Studies* 18(1): 3–21.

Rosson, C. P., C. E. Curtis, G. W. Auburn, and G. L. Carriker. 1989. "The International Soybean Situation: Can Southern Producers Compete?" *Journal of Agribusiness* 7(1): 49–55.

Schuh, G. Edward. 1987. "The Changing Context of Food and Agriculture Development Policy." In *Food Policy, Integrating Supply, Distribution, and Consumption,* ed. J. Price Gittinger, Joanne Leslie, and Caroline Hoisington, pp. 72–88. Baltimore: Johns Hopkins University Press.

Selden, Mark. 1993. "Beyond Collectivization: Socialist and Post-Socialist Agrarian Alternatives in Asia and Europe." Paper presented at the Future of East Asian Socialism Conference, Canberra, Australia, January.

Simons, Marlise. 1993. "A Dutch Reversal: Letting the Sea Back In." *New York Times,* March 1: A12.

Sklair, Leslie. 1991. *Sociology of the Global System.* Baltimore: Johns Hopkins University Press.

Sorj, Bernardo, and John Wilkinson. 1990. "From Peasant to Citizen: Technological Change and Social Transformation in Developing Countries." *International Social Science Journal* 124: 125–33.

Spitz, Pierre. 1985. "The Right to Food in Historical Perspective." *Food Policy* 10(4): 306–16.

Talbot, John. 1994. "The Regulation of the World Coffee Market: Tropical Commodities and the Limits of Globalization." In *Food and Agrarian Orders in the World-Economy,* ed. Philip McMichael. Westport, Conn.: Greenwood Press.

Thurow, Lester. 1992. *Head to Head: The Coming Economic Battle among Japan, Europe, and America.* New York: William Morrow.

Tilly, Charles. 1975. "Food Supply and Public Order in Modern Europe." In *The Formation of National States in Western Europe,* ed. Charles Tilly, pp. 380–455. Princeton: Princeton University Press.

Tomich, Dale. 1990. *Slavery in the Circuit of Sugar: Martinique and the World Economy.* Baltimore: Johns Hopkins University Press.

Ufkes, Frances. 1993. "Trade Liberalization, Agro-Food Politics, and the Globalization of Agriculture." *Political Geography* 12(3): 215–31.

UN Conference on Trade and Development (UNCTAD). 1990. *Agricultural Trade Liberalization in the Uruguay Round: Implications for Developing Countries.* New York: United Nations.

U.S. Department of Agriculture. 1992a. *Former USSR. Agriculture and Trade Report.* Situation and Outlook Series. Washington, D.C.: USDA-ERS.

——. 1992b. *China. Agricultural and Trade Report.* Situation and Outlook Series. Washington, D.C.: USDA-ERS.

Vocke, Gary. 1987. "The Green Revolution Lags Rising Wheat Consumption in the Developing World." *World Agriculture. Situation and Outlook Report,* pp. 10–19. Washington, D.C.: USDA-ERS.

Vollrath, Thomas, and Linda Scott. 1991. "Developing Countries as a Source of U.S. Export Growth." In *Developing Economies Agriculture and Trade Report,* pp. 16–26. Washington, D.C.: USDA-ERS.

Wallerstein, Immanuel. 1974. *The Modern World-System: Capitalist Agriculture and the Origins of the World Economy in the Sixteenth Century.* New York: Academic Press.

——. 1992. "The Collapse of Liberalism." In *Socialist Register,* ed. Ralph Miliband and Leo Panitch, pp. 97–110. London: Merlin Press.

Walton, John. 1990. "Debt, Protest, and the State in Latin America." In *Power and Popular Protest: Latin American Social Movements,* ed. Susan Eckstein, pp. 299–328. Berkeley: University of California Press.

Watkins, Kevin. 1991. "Agriculture and Food Security in the GATT Uruguay Round." *Review of African Political Economy* 50: 38–50.

Whatmore, Sarah, Richard Munton, and Terry Marsden. 1991. "Research Developments on Rural Change: A British Perspective." Paper presented to the American Sociological Association annual meeting, Cincinnati, Ohio, August.

Winters, L. A. 1990. "The Road to Uruguay." *Economic Journal* 100(403): 1288–303.

Wolf, Eric. 1966. *Peasants*. Englewood Cliffs, N.J.: Prentice Hall.

——. 1982. *Europe and the People without History*. Berkeley: University of California Press.

Index

1740

Food Systems and Agrarian Change

Edited by Frederick H. Buttel, Billie R. DeWalt,
and Per Pinstrup-Andersen

Hungry Dreams: The Failure of Food Policy in Revolutionary Nicaragua, 1979–1990
by Brizio N. Biondi-Morra

Research and Productivity in Asian Agriculture
by Robert E. Evenson and Carl E. Pray

The Politics of Food in Mexico: State Power and Social Mobilization
by Jonathan Fox

Searching for Rural Development: Labor Migration and Employment in Mexico
by Merilee S. Grindle

The Global Restructuring of Agro-Food Systems
edited by Philip McMichael

Structural Change and Small-Farm Agriculture in Northwest Portugal
by Eric Monke et al.

Diversity, Farmer Knowledge, and Sustainability
edited by Joyce Lewinger Moock and Robert E. Rhoades

Networking in International Agricultural Research
by Donald L. Plucknett, Nigel J. H. Smith, and Selcuk Ozgediz

Adjusting to Policy Failure in African Economies
edited by David E. Sahn

The New Economics of India's Green Revolution: Income and Employment Diffusion in Uttar Pradesh
by Rita Sharma and Thomas T. Poleman

Agriculture and the State: Growth, Employment, and Poverty in Developing Countries
edited by C. Peter Timmer

The Greening of Agricultural Policy in Industrial Societies: Swedish Reforms in Comparative Perspective
by David Vail, Knut Per Hasund, and Lars Drake

Transforming Agriculture in Taiwan: The Experience of the Joint Commission on Rural Reconstruction
by Joseph A. Yager